James C. Bayles

House Drainage and Water Service in Cities, Villages, and Rural Neighborhoods

With incidental consideration of causes affecting the healthfulness of dwellings

James C. Bayles

House Drainage and Water Service in Cities, Villages, and Rural Neighborhoods
With incidental consideration of causes affecting the healthfulness of dwellings

ISBN/EAN: 9783337377267

Printed in Europe, USA, Canada, Australia, Japan

Cover: Foto ©berggeist007 / pixelio.de

More available books at **www.hansebooks.com**

HOUSE DRAINAGE

AND

WATER SERVICE

IN CITIES, VILLAGES, AND RURAL NEIGHBORHOODS.

WITH INCIDENTAL CONSIDERATION OF CAUSES AFFECTING
THE HEALTHFULNESS OF DWELLINGS.

BY

JAMES C. BAYLES,

EDITOR OF "THE IRON AGE" AND "THE METAL WORKER."

———•———

NEW YORK:
PUBLISHED BY DAVID WILLIAMS, 83 READE STREET.
1879.

CONTENTS.

CHAPTER I.
Hygiene in its Practical Relations to Health . . . 5

CHAPTER II.
Sewer Gas 23

CHAPTER III.
Waste and Soil Pipes 44

CHAPTER IV.
Traps and Seals and the Ventilation of Soil Pipes . . 64

CHAPTER V.
Water-Closets 85

CHAPTER VI.
Service Pipes and Water Service in City Houses . . 104

CHAPTER VII.
Tanks and Cisterns 139

CHAPTER VIII.
The Chemistry of Plumbing 147

CHAPTER IX.
Elementary Hydraulics Applicable to Plumbing Work . 216

CHAPTER X.
Sanitary Construction and Drainage of Country Houses . 258

CHAPTER XI.
Water Supply in Country Districts . . . 291

CHAPTER XII.
Suggestions Concerning the Sanitary Care of Premises 312

CHAPTER XIII.
The Plumber and His Work 328

AUTHOR'S PREFACE.

For several years the writer has conducted an extensive correspondence with plumbers, builders, architects and others interested in the mechanics of hygiene, growing out of the discussion in *The Metal Worker* of practical questions pertaining to plumbing and sanitary engineering. From this correspondence, as well as from a careful study of the literature of the subject, he learned that a need existed for a comprehensive elementary treatise on the theory and practice of plumbing which was not met by any work in the market. The idea of an attempt to supply this want did not at first suggest itself, however, as continuous and exacting professional engagements seemed to leave no time for book-making. During the winter of 1874-5 the writer had the honor of reading two papers before the Public Health Association of New York, both on topics connected with house drainage, which were so favorably received by the plumbing trade as to suggest the propriety of revising them for republication in pamphlet form. The work thus begun gradually expanded, until it assumed the proportions of a book, and it was then deemed advisable to still further extend its scope to include the whole subject of house drainage and water service.

It is perhaps only just to the professional reader to say that this book is not intended as a contribution to the literature of sanitary engineering. It takes up the subjects of drainage and water supply where the engineer commonly leaves them, and treats almost exclusively of subjects in which householders and those connected with the house-building trades are directly and immediately interested. On the other hand, it does not claim to be a workshop manual. There is little in the simple manipulations of the plumber's art to call for explanation. The plumbers as a class need theoretical instruction chiefly, and this is equally valuable to the other classes of readers addressed in these pages, namely, architects, builders, householders and physicians interested in studying the mechanics of hygiene. The writer has learned from experience that to be of value such a work must be at once elementary and thorough. It should aim to supply exactly the information which the reader is likely to find practically useful; and while the discussion needs to be full and exhaustive, it is of little use to encumber the pages with citations from foreign authors or with many references to works not readily accessible to the general reader. With this in view the writer has been somewhat more didactic in his treatment of the subjects discussed than he would have been if writing for experts in hygienic science; and the book is published in the hope that it will be of value to the readers specifically addressed and aid in creating a popular interest in matters intimately affecting the public health.

The author desires to acknowledge his obligations to those who have assisted him in various ways, especially to Mr. William E. Partridge for valuable aid in nearly every line of investigation and experiment; to Messrs. J. W. Hallock and B. C. Gregory for assistance in chemical research and laboratory work; to Mr. John Birkinbine, C. E., for aid in investigations pertaining to the science of hydraulics; to Prof. Charles F. Chandler for documents and information; to Drs. George Bayles and Elisha Harris for valuable data; to Mr. William Emerson for exceptionally careful and intelligent proof reading, and to many kind friends in nearly all parts of the United States who have aided his work and encouraged him in completing it for publication.

83 READE STREET, NEW YORK, April, 1878.

CHAPTER I.

Hygiene in its Practical Relations to Health.

It is a gratifying indication of the progress of civilization that sanitary science, as it is called, is becoming, even to a limited extent, a popular study. *Sanitary science becoming a popular study.*

For many centuries physicians had a practical monopoly of what little was known of the conditions affecting the public health, and there seemed to be no incentive to original investigation and experiment, even if the means of prosecuting an inquiry so important to all classes of the people had been at the command of those who, under more favorable circumstances, would, doubtless, have made important contributions to the literature of hygiene. Fortunately, the science of medicine—if that could be called a science which was then empirical, and still is to a great extent—gradually freed itself from the hideous superstitions which so long trammeled it, and physicians began to open their eyes to the real teachings of experience, and to treat disease rationally. This was a great step forward. The next step was to push the inquiry into the causes of disease, and the means by which those causes could be reached and extirpated—or, at least, so far controlled as to essentially modify their power for mischief. To the medical profession we owe the greater part of what has already been learned and placed upon record, of the truths which form the basis of sanitary science; but though long left to pursue their studies without encouragement, and with little or no hope that even their most startling discoveries would be appreciated by the general public, they have at last drawn to their assistance in the great work a large and influential class. Those to whom the sanitarian must look for the practical application of his carefully elaborated theories of sanitary reform, are the very ones who *The beginnings of sanitary investigation.* *Services of the medical profession.*

6 HYGIENE IN ITS PRACTICAL RELATIONS TO HEALTH.

Popular interest in sanitary reform. are now beginning to appreciate most fully the importance of sanitary science. In all classes of society, except those which include the very degraded and ignorant, we find a growing interest in the means of guarding against all unhealthful conditions in person, house and environments. With a larger wisdom and clearer insight into the causes of things, which have come of progress in scientific thought, intelligent people do not *The true causes of disease.* now attribute the consequences of their own neglect and carelessness to the "afflictive dispensations of Providence," which are "mysterious and past finding out." We are beginning to understand how large a proportion of the diseases which afflict humanity results from preventable causes, and that it is possible, by judicious measures of sanitary reform, to so reduce the death rate as to materially increase the average duration of human *The practical benevolence of sanitary work.* life. Nor is this interest in sanitary reform bounded and limited by a narrow selfishness. There is something broadly humanitarian in it. The rich and middle classes no longer feel that they have no interest in the welfare and comfort of those who endure the misery and utter wretchedness of squalid poverty. Disease is no respecter of persons, and a "fever nest" in some remote and neglected quarter of a populous city may dispatch invisible messengers of death to poison the air of broad avenues and clean-swept streets miles distant. The enlightened self-interest which is leading so many intelligent men and women to study the laws of health is exactly the reverse of selfishness, since every movement for general sanitary reform begins with the improvement of the houses of the poor and ignorant, who can only be redeemed from untimely death, and saved from being the instruments of spreading the seeds of disease and contagion, when those who occupy the social planes above them stretch forth a helping hand to lift them out of the mire into which they have fallen.

The task of the sanitarian not hope When we look about us and see how much remains to be done before the masses of the people shall be emancipated from the dire necessity of living under conditions prejudicial to health and, consequently, to happiness, it would almost seem as if the task of the sanitarian is a hopeless one. Such is not the case, however. We turn with a shudder of horror from the records of the past to contemplate with satisfaction the progress we have already made. People often wonder why we do not have such fearful visitations of epidemic at the present day as the plague of London, the ancient spotted fevers, sweating sickness, &c. They forget that we are not yet free from the cholera, the yellow fever, typhoid fever, and other preventable diseases, and that the next generation may see that our disregard of nature's laws affected our death rate as surely as the dirt and filth of London caused the great plague. *Epidemics.*

From the fall of the Roman Empire to the end of the Middle Ages, the people of Europe were unwashed. Of Paris, it is recorded by Rigord, physician to Philip Augustus, that one day when the king, walking to and fro in his audience chamber, went to look out upon the view for recreation, some carriages belonging to citizens happened to pass in the street beneath the window, "when the substance forming the street, being stirred up by the revolution of the wheels, emitted a stench so powerful as to overpower Philip. This so disgusted the king that he urged the citizens to pave the streets, and, to assist in effecting the purification of the city, he built a wall around the cathedral to prevent it from remaining longer a common corner of convenience." These measures occasioned great popular dissatisfaction, and we really have no reason to wonder that plagues and pestilences were so common in a city with such streets, and in which the angles of the cathedral walls were used as privies. One writer, in speaking of the condition of London about this time, says that in the streets around St. Paul's Churchyard the "horse manure was a yard deep," and also speaks of the streets as never having been cleaned. Public muck heaps were found at every corner. "Floors were of clay covered with rushes which grew in the fens, which were so slightly removed now and then that the *Life in Europe in the Middle Ages. The streets of Paris in the 12th century. London in the 12th century.*

lower part remained sometimes for twenty years together," and in it such a collection of foulness as we should expect to find only in a city scavenger's cart. The chronicle goes on to specify of what the filth consisted, but I omit the items for the sake of decency. The odors were horrible, and to disguise them perfumes were largely used and fragrant gums were burned to sweeten the air. Cleanliness of person was a thing almost unknown. One old chronicler says of the ladies: "They wore clean garments on the outside, but the dirty ones were often worn until they fell away piecemeal from their unwashed bodies." The history of Chester shows the fearful effects resulting from the utter neglect of sanitary precautions which seemed to be characteristic of our English ancestors. I quote as follows: "In 1507 sweating sickness was very severe in Chester for three days; 91 died. In 1517, great plague; grass a foot high in the streets. 1550, sweating sickness. 1603, great plague began in one Glover's house, in which 7 persons died; 60 died weekly, in all 650 persons, and 61 of other diseases. 1604, plague; very hot; 812 deaths. 1605, plague still increasing; 1313 died of it, beside those of different diseases." In 1649, 2099 persons died of the plague. And so the record goes. The people prayed for deliverance from sickness and death, but forgot their garbage heaps, their foul streets, dirty houses and personal uncleanliness.

The gaol fever was another disease which was much dreaded by all classes of the people, and its ravages show how utterly sanitary precautions were neglected in prisons. Unfortunates and criminals were confined in damp, cold, unventilated cells, and kept in a state of inactivity, without a chance for fresh air or exercise. The stench from their own bodies and the absence of any means of purifying their persons, bedding and clothes during confinement, filled the air with exhalations so poisonous that sickness was inevitable. The prison house became a prolific source of contagion, and though the prisoner might escape death, he carried in his clothes, when liberated, the seeds of

HYGIENE IN ITS PRACTICAL RELATIONS TO HEALTH. 9

sickness and death to others. The Black Assize at Oxford, in 1577, is a memorable event which serves to show us, by contrast with criminal court terms of the present day, what progress has been made since that time. Baker tells us, in his Chronicle, that all who were present in the court died of gaol fever within forty-eight hours—judge, lawyers, constables, witnesses, prisoners and spectators—in all some three hundred persons. In London the great plague would have been a matter of annual recurrence, and the hundred thousand who died of it would have been only the advance guard of the army of victims, had not the great conflagration, which soon followed it, purified the city with fire. When it was rebuilt more attention was given to sanitary laws, which were just beginning to be understood, and the new city, being comparatively clean, escaped the contagion which loaded the air of the old. *The Black Assize. London plagues.*

When we have in mind such facts as these, gathered at random from the annals of past centuries, we realize that all classes of society share—though not in equal degree, perhaps—the benefits of the steady upward progress toward higher standards of civilization and social refinement. In Europe and America we see the growth of societies of thoughtful, earnest men, organized to discuss questions affecting the public health, and to devise means of making unthinking and unthankful communities healthier and happier. In many cities we see liberal appropriations of public money expended by boards and commissions composed of men eminent for scientific attainments and public spirit in sanitary work, while an army of self-sacrificing physicians labor in the work of sanitary inspection with a zeal and fidelity to duty altogether disproportionate to their scant remuneration—if the value of such services can be measured in money. We see the steady and sustained progress of improvement in the comfort, convenience and healthfulness of the homes of the upper and middle classes, and we also see representatives of these classes devoting time and means to further the great work of bettering the condition of those below them *The past and the present. Sanitary administration in cities. Public sanitary work.*

in the social scale. Such associations as the Artisans', Laborers' and General Dwellings Company, of London, which has built the workingmen's city at Shaftesbury Park, and the Dwellings Reform Association, of New York, having for its object the provision of better, more commodious and more wholesome homes for the neglected poor now crowded into foul and dirty tenements, are the outgrowths of an enlightened and liberal public sentiment, and the operations undertaken and proposed by them would be impossible of accomplishment under any other conditions than those which exist in London and New York. It is so in many things. Progress in civilization has given us hospitals and dispensaries for the sick, built asylums for the insane, and provided clothing, food and shelter for the pauper, organized and carried out great schemes for the relief of suffering, and in innumerable ways extended its benefits to those who contribute least to it. Society recognizes its duty and honestly, though not always wisely, seeks to perform it. Public sanitary work is a part of this great scheme—one of the fruits of practical Christianity in highly civilized communities, and the sanitarian who seeks to extend the knowledge and promote the intelligent study of Nature's laws, renders important service in the cause of human progress.

Importance of a popular understanding of Nature's laws. But while sanitary science is beginning to attract some share of public attention, the reforms and improvements which it seeks to effect in the conditions of our everyday life are not easily accomplished. Much has already been done in this country, and more in England, in devising and carrying out systems of sanitary reform, but the truths upon which sanitary science is founded must be deeply impressed upon the public mind before we can look for great and important results. This popular education can only be accomplished gradually by the patient and intelligent teachings of unselfish specialists through the medium of the newspapers, in books, in pamphlets and tracts, presenting elementary truths in such shape as to command attention for them, and through the work of such societies as the American Public Health Association, the Public Health Association of New York, and similar organizations. Little of either fame or profit can be expected to result from this preliminary work in the field of sanitary reform, but those who engage in it with honest and unselfish purpose do not, as the rule, desire other reward than the knowledge that they are doing something for the good of humanity.

So far as regards the movement looking to the reform of the evils to which the reader's attention is directed in the succeeding chapters, its success depends very much upon our architects. When they call for good plumbing work in their specifications, knowing what they want and refusing to accept anything else, they will have no difficulty in getting it. When capitalists are willing to pay the price of good work, the architects will learn what good work is and how to call for it. In most other respects our architecture is very well adapted to our climate, our social life and our present needs. As a people, we live in more comfortable houses than are found in any other country of the world. None appreciate this so fully as those who have traveled observingly in foreign countries and studied the home life of other peoples. Our dwellings of the better class are finished and fitted up with a completeness and a regard to comfort and convenience which astonishes foreign architects. In the sundry items classified under the general name of "modern conveniences," our architectural practice has fairly kept pace with the development of the various industries connected with the building trades; and even in the dwellings of the middle classes we find evidence of an intelligent regard for the comfort of the occupants not seen in dwellings of the same class in any part of Europe.

There is a reason for this. During the brief period of our national life the building trades have necessarily been among the most important of our great national industries. To provide homes for our rapidly-growing population, we have been compelled to build more houses than have probably been built in all Europe during the same time. We are, moreover, a home-loving and an inventive people, and have given a generous encouragement to well-directed efforts to improve our house fixtures. A glance over the annual reports of the Patent Office at Washington will show that a very large percentage of the inventions patented are labor-saving appliances, designed to find a place in the domestic economy. Generally speaking, we have, as a people, very sensible ideas of comfort, and are not much hampered by either custom or precedent in these matters. We do not, like the conservative Englishman, retain the open fireplace because of its traditions and from a mistaken notion that comfort and health are incompatible in house-warming. We discarded the open fire a generation ago, and adopted the more economical and efficient iron stove; now the stove is giving place to the hot-air furnace, and this, in turn, will be pushed aside by the steam heater in first-class work. This restless desire for improvement has kept the inventive talent of the nation directed to the changing requirements of the building trades, and has enabled us to attain, even in cheap construction, a degree of comfort which in other countries would be deemed extravagant luxury. On this score, at least, we have no just quarrel with the architects.

<small>Home comforts.</small>

<small>Warming houses.</small>

But while convenience and comfort are certainly desirable in an eminent degree, they are not the only qualities to be sought in house building. These we demand, and properly; but out of the limitations which those who build houses and those who buy them have fixed to the intimacy of the relations of science and art to architectural practice, have grown other and very serious evils. We may divide these evils into two general classes—those which are just beginning to attract the attention of the hygienic physicist, and those which have long received the thoughtful consideration of the economist.

<small>Evils in architectural practice.</small>

In the first of these general classifications we may include the evils inevitably attendant upon a disregard of hygienic laws in house building; in the second are included subjects which cannot properly claim consideration in these pages.

HYGIENE IN ITS PRACTICAL RELATIONS TO HEALTH. 13

It is a fact which unfortunately does not admit of intelligent contradiction, that in the architectural practice of the time very little attention is paid to the laws of health. What is known as sanitary science is still to some extent empirical; but from experience we have learned something of Nature's laws and Nature's penalties, and we certainly have a right to expect that our architects shall not, by disregarding the former, force us to incur the latter. Let us begin with our heating apparatus, already noticed as contrasting so favorably, on the score of comfort, with the primitive fireplace of Great Britain and the clumsy, inefficient appliances employed on the Continent. Owing to the length and severity of our winter seasons, the furnace is one of the most important of the permanent fixtures of a well-appointed house. Now, it is by no means probable that the system of heating by the distribution of air currents moderately warmed by contact with the radiating surfaces of a furnace, is objectionable on hygienic grounds. It is the abuses of the system which give rise to the evils commonly charged against the system itself, and in these abuses we find a marked difference between scientific theory and every-day practice in architecture. It is probable that every well-informed architect is familiar with the fact that there is a vast difference, as regards its healthfulness, between a system of heating in which a large volume of moderately-heated air is employed and one in which dependence is placed upon a small volume of air raised to a high temperature. The very common abuse of the system consists, principally, in the use of furnaces too small for the work they have to do. As the consequence, we must drive them in cold weather to such an extent that the air passing through them is vitiated and rendered unfit for breathing. We cannot expect the average householder to understand these matters, and we must look to the architect to lead the progress of reform which shall give us wholesome heating without sacrifice of comfort.

Intimately connected with the problem of healthful warming,

14   HYGIENE IN ITS PRACTICAL RELATIONS TO HEALTH.

is that of ventilation. Here the difference between theory and practice in house building—between what we know should be done and what we do or attempt to do—is certainly very marked. The subject of ventilation has a voluminous literature of its own, with which the well-read architect cannot but be more or less familiar. Probably he appreciates more fully than any one but the specialist in practical hygiene, the importance of good ventilation in dwellings; but in not one in a hundred of the dwellings he builds is any provision whatever made for ventilation. What is simple and comparatively easy of accomplishment at the hands of an intelligent architect when he plans a building, becomes difficult and often practically impossible of accomplishment after the house is finished, without costly and troublesome reconstruction. That the average architect is practically ignorant of the mechanical means by which adequate ventilation can be secured in cold climates without unnecessary waste of fuel, is no more to be wondered at than that he so often fails in his essays in the domain of high art. With us it is not yet a part of the business of house making, and we do not give him an opportunity to learn from practical trial the fact that, to secure good ventilation, it is only necessary to remove impure air, and that, with the whole volume of the atmosphere exerting on all sides a pressure equal to about 14 pounds to the square inch, it is as idle to pump fresh air into a building as it is to pump water down hill. Hence, when we call upon the architect of average skill to exercise the functions of an engineer of ventilation, he is more likely to fail than to succeed. We see this illustrated in the bad ventilation of our churches and public halls—if that may be called ventilation which does not ventilate—and if we pursue the experiment long enough, and without regard to expense, we are likely to reach results almost as unsatisfactory as those secured in the effort to ventilate the House of Representatives at Washington. We blame the architect for the impure air of our dwellings and places of assembly, but when he undertakes

Ventilation of dwellings.

Artificial supplies of fresh air.

Bad air.

to give us good ventilation and fails, all he is really to blame for is over-confidence in essaying a task for which he has neither the education nor the experience. In such a climate as we have in New York, we cannot have both economical heating and good ventilation unless we build our walls and floors with non-conducting filling. As we do build, however, we are content to do without the ventilation; and, to secure both comfort and fuel economy, even the scanty supply of fresh air which comes in around our doors and window sashes we cut off in the early autumn with list and weather-strips. We are not only content to do without ventilation, but we positively do not want it in any form in which it has yet been given to us. Some years ago a wealthy and philanthropic land owner in one of our principal cities, conceived the idea of erecting a number of healthy houses which should be built on scientific principles. Ventilation was especially sought, and the best talent at command was engaged to provide the necessary appliances; but when the houses were finished the owner found himself unable to retain his tenants except upon the condition that he would seal all his ventilators. Probably the tenants were not so blind to their own interests as might appear at first glance. No doubt it was impossible to keep these houses warm enough for comfort, owing to the loss of heat by absorption into the walls and its escape through the ventilators. In ventilation, comfort and health are almost synonymous, and when we can have the benefits of pure air without a ruinous consumption of fuel or the discomfort of low temperatures, we shall no longer object to it; indeed, we shall demand it.

Fuel economy

Popular indifference to ventilation.

That the educated architect should thoroughly understand the principles and the methods of ventilation, is too obvious to need the support of argument. It is not, however, an art which can be acquired easily or from mere generalizations. Nor will it help him much to master the details of a "system," however good that system may be, for the reason that no system can be devised which will admit of successful application under various conditions. A system which would work well in one house might fail in part in another house, and fail utterly in a public hall; while a system applicable to a church or a lecture room would probably be little better than no system at all in a theater or hospital. There are, however, certain principles which apply to the ventilation of all classes of buildings which are so simple and, when learned, so obvious, that the architect rarely attempts to apply them until he has tried all other plans unsuccessfully. It is a curious fact that those who give attention to ventilation rarely avail themselves of the experiences of their predecessors. Beginning where they began, they go through pretty much the same course of trials and failures, and it is generally an easy matter to tell how much experience a man has had by ascertaining what "system" he tried last. When the importance of good ventilation is better understood by the public, and the architect is required to provide it in our dwellings, he will probably find it to his interest to call to his aid the specialist who has made ventilation his study, and who has learned from experience how to meet all the conditions which complicate the problem so seriously.

Mistakes of architects.

In the defects found in the average plumbing work of the time, we see another instance of the wide difference which exists between the measure of our scientific knowledge and the methods of our architectural practice. No fact rests upon a broader and more substantial basis of truth than that the gaseous emanations from decomposing sewage, commonly called sewer gas, are a fruitful source of disease. Whatever the agency by which sewer gas works, we know that it comes armed with the power and potency of death. Escaping into the free atmosphere, its deadly power is quickly destroyed by the oxidation of its organic poisons; but when it mingles with the confined air of our unventilated living and sleeping rooms, it retains its terrible power for mischief long enough to do its deadly work effectually. Dr. Mapother, of Dublin, an eminent authority, states that there occur annually in England 140,000

Plumbing work.

Sewer gas in dwellings.

cases of typhoid fever, of which 20,000 terminate fatally, which are clearly traceable to defective drainage and sewer-gas poisoning, and yet typhoid fever is only one of a long list of prevalent zymotic diseases. England and Scotland together gave in the five years ended January 1, 1870, deaths from zymotic diseases amounting to 21·9 of the total mortality, as shown in returns made by order of Parliament in 1871. The variation of the zymotic ratio in the sum of causes of mortality ranges from 10 to 37 per cent. of the total deaths. From such imperfect statistics as have been gathered in this country, it is safe to conclude that zymotic diseases cause, directly or indirectly, about one-half the deaths occurring in our great cities. In the vital statistics of New York for the past 11 years, zymotic diseases, as now classified, are charged with about 32 per cent. of the deaths from all causes. The figures are as follows:

Typhoid fever in England and Scotland.

Mortality from zymotic diseases.

In New York.

New York City.	Deaths from zymotic diseases.	Percentage of total mortality.
1866	8,788	32·77
1867	6,583	28·41
1868	7,456	29·06
1869	7,676	30·50
1870	8,314	30·60
1871	8,964	31·01
1872	11,815	36·19
1873	9,505	32·98
1874	9,715	33·82
1875	10,964	35·52
1876	8,538	29·25

In some of our principal cities the percentage is higher than in New York. In others it is much lower, as will be seen from the following comparison of the average ratio of deaths from zymotic causes to the total annual mortality:

Other American cities.

Pittsburgh.......................... 35 per cent.
Chicago............................ 34 "

Brooklyn	33 per cent.
Boston	33 "
Cincinnati	33 "
Milwaukee	31 "
Baltimore	28 "
Washington	25 "
San Francisco	22 "
Philadelphia	20 "

If it be assumed that the relation of deaths to the number of cases of sickness induced by zymotic causes is about the same here as in the case of typhoid fever in England, the effect of filth poisoning upon the public health will with difficulty be realized. If we look for the cause of this large mortality from diseases of the zymotic type in our cities, we find it principally in sewer-gas poisoning. Other causes operate to swell the total, but to bad plumbing work we may attribute the prevalence of pythogenic pneumonia, peritonitis, inflammatory rheumatism, typhoid and malarial fevers, croup, diphtheria and many kindred diseases which are almost epidemic in our large cities.

Filth poisoning.

Unfortunately for the progress of hygienic reform, the difference between good and bad plumbing work is usually so slight as to escape the notice of any but the trained expert; but it is commonly great enough to exert an active and far-reaching power for mischief. We expect to find in the houses among which we seek homes for our families all the conveniences which are rendered possible by the vast systems of hydraulic engineering which find their consummation in the water service and drainage of a city house. The bath, the water-closet, stationary wash basins with hot and cold water, laundry tubs, the butler's pantry and the kitchen water system, are no longer regarded as luxuries but as necessities in all well-appointed modern houses. There is no good reason why we should not have all these conveniences, but we often pay a fearful price for them. Let us follow the intelligent sanitary inspector in an examination of the pipe systems of an average New York house of the better class.

The difference between good and bad plumbing.

Fixtures in houses.

HYGIENE IN ITS PRACTICAL RELATIONS TO HEALTH.

Beginning with the water service, we find that the pipes are of lead, notwithstanding the fact that the architect has ready to his hand several kinds of pipe quite as convenient as lead and much safer than those made of a metal which, under a great variety of conditions, parts with poisonous salts to the water passing through it. All conscientious architects familiar with the literature of chemistry will admit that lead should be discarded as an unsafe metal for service pipes, and tin or black iron used instead; but lead is still called for in ninety-nine out of every hundred specifications. In the drainage system and its appurtenances we find evils of a different and more serious character. We see dependence for the suppression of gases, often held under considerable pressures in the sewers, placed upon supposititious half-inch water seals in traps of such shape and so placed that they are likely to be emptied, from one cause or another, every hour in the day, and to stand empty at night. We find that the foul sewer is provided with breathing holes into our houses; that in dark, unventilated recesses adjoining our bedrooms are cheap and flimsy water-closets, wrong in principle and wholly unsatisfactory in operation, which retain so much of the filth passing into them that they become pestilent nuisances. In short, we find every condition so favorable to sewer-gas poisoning that we no longer wonder at the great mortality from diseases of pythogenic origin in our sewer-drained cities. As the plumbing work of our houses is commonly done, it would be better for most of us if we had to bring our water in buckets from a public hydrant, and carry our waste to the culvert at the nearest street corner.

Where shall we place the responsibility for this most terrible of the evils which characterize the architectural practice of the time? We know from experience that very few of our architects have given the problems of hygienic house drainage the careful attention they deserve, but it is not because they do not know the consequences of cheap and defective plumbing work in houses, nor because they consider these defects irremediable.

[marginalia: Service pipes. Drainage. Traps. Water-closets. The responsibility for bad plumbing.]

The evils to which we have called attention exist and multiply, simply because the architect in general practice cannot insist upon a due observance of hygienic laws in house construction and compete successfully with those in the profession who are *The architect.* less conscientious in these matters. If his clients neither know nor care whether a house is well or badly drained, why should he drive away business by demanding that we shall pay for good plumbing work, when others will furnish us equally acceptable plans and specifications which can be followed in construction more cheaply? Consequently, the architect rarely troubles himself to learn the theory of plumbing, save in the most superficial way. His specifications of pipes and fixtures are usually so loosely drawn as to be susceptible of the most liberal interpretation by those who bid upon them. As the lowest bidder commonly secures the contract, we may be sure that every advantage will be taken of the incompleteness and ambiguity of *Specifications.* the specifications, which are rarely specific except as to the number and kind of fixtures to be supplied and the weight of lead pipe to be used. The shrewd, practical plumber knows just how much regard it is necessary to pay to the stereotyped phrases which provide that his task shall be performed "in a workmanlike manner, and to the satisfaction of the architect and owner." The architect gives the work only a cursory supervision at most, and the owner is commonly satisfied if the fixtures are all in the right places and look as he expected. *The owner.* A stain in a marble slab or a thin spot in the silver-plating of a basin cock is far more likely to give dissatisfaction than a soil pipe of paper thickness, put together with mason's cement or glazier's putty, instead of substantial pipe weighing (if of 4 inches diameter) not less than 12 pounds to the foot, and put together with well-calked lead joints.

The specialist in the field of practical hygiene naturally blames the architect for the existence of evils so prejudicial to *A divided responsibility* the public health; but there is a divided responsibility. The architect shifts his share upon the builder, the builder upon the parsimonious owner unwilling to pay the price of good work, and the owner upon the "rascally plumber" who "scamped the job." But it does not rest here. The plumber replies that he works for a profit, and means to make it when he can. If the owner expected to get a thousand dollars' worth of materials and time for five hundred dollars, he is the only party to the transaction who is deceived, and that because he deceived himself. There is something of truth in each of these specious disclaimers, but perhaps the architect has a larger share of the moral responsibility than he is willing to admit. If he would let discreditable work go to those more anxious for present gain than for an honorable professional reputation, we should be better able than we now are to draw the line between the two classes composing the profession. *The plumber.*

It is, perhaps, too much to expect that there will ever be in our average architectural practice a close approximation to the measure of our scientific knowledge. If it follows, even a long way behind, the footprints of invention and discovery, it will be as rapidly progressive as we can hope to see it. Generally speaking, we gain knowledge a good deal faster than we can practically apply it, and our progress toward higher standards in architecture will and should be characterized by a judicious conservatism. The material interests involved are large, and must be carefully guarded by the conscientious architect. We cannot, therefore, expect that he will make haste to utilize every new fact which may be added from day to day to the sum of the world's knowledge, but we have a right to insist that he shall not carry his conservatism too far, and cling to systems and methods entailing evils from which we naturally and properly look to him for protection. In these matters there should be a much closer relation than now exists between theory and practice in architecture, and if the conscientious architect will first educate himself in those branches of his art in which the disparity is greatest, he will find it an easy task to bring about the desired reforms. In thus educating the public, *Conservatism in architecture.*

by placing before them the results of his own education, he will open for himself a broader and nobler field of usefulness, in which he will be less hampered by the limitations and restrictions of which he now complains.

Good plumbing obtainable when wanted. As for the plumber, I can say with confidence that, so far as new work is concerned, he will give us what our architects call for, and when he has only good work to do in new buildings, he will soon learn what good work is and how to avoid mistakes in jobbing.

CHAPTER II.

SEWER GAS.

Popular indifference to the evils resulting from defective drainage is, doubtless, attributable wholly to popular ignorance. A majority even of intelligent people regard the subject as one in which they have no personal interest, and for this reason it is difficult to instruct them through the medium of the public prints. Those who have been engaged in the work of sanitary inspection have almost invariably experienced great difficulty in securing reforms, even of the most dangerous evils; and, unless supported by legal authority, their suggestions and directions are nearly always disregarded. The popular belief seems to be that there is a great deal more talk about sewer gas among those who lay claim to scientific knowledge, than its practical importance really warrants. I scarcely need assure the reader that this is a serious mistake, which cannot fail to imperil the public health by giving rise to a false sense of security and encouraging the toleration of dangerous nuisances. In many respects the ancients were wiser in sanitary matters than the moderns. No nation ever had a code of laws embodying so much of sound, practical wisdom—so far as regards hygiene, at least—as the Jews under Moses and his immediate successors, and the more we learn of Nature's laws, the better we understand and appreciate the importance of the regulations established for the government of the tribes of Israel in their long journeying after the exodus from Egypt. *Difficulty of securing sanitary reforms. Popular errors. Sanitary laws of the Jews.*

When architecture reached its highest æsthetic development, and drainage systems were adopted, the importance of guarding against the danger of sewer-gas poisoning seems to have been well understood, for the ruins of ancient Rome show that all the *cloaca* were well ventilated, to the end that the pure atmos- *Sewer ventilation in ancient Rome.*

pheric air might oxidize and destroy the poisons arising in the gases given off by decomposing sewage. The knowledge which prompted these precautions has never been lost to the world, but for some reason which it would be difficult to explain, modern engineers and architects have too generally neglected the simple precautions so necessary to the protection of the public health, and, as the rule, modern sewers are but indifferently ventilated, if at all. As a consequence, the gases generated in our sewers are rarely rendered innoxious by dilution with enough pure air to destroy the organic germs which go with them, and when they find their way into a house they are pretty sure to cause serious mischief.

<small>Modern sewers not generally ventilated.</small>

What is sewer gas? The most careful analyses show that it is composed chiefly of carbonic acid, nitrogen, sulphureted hydrogen, ammoniacal compounds and fœtid organic vapor. The elementary gases and those of known composition, which are commonly found in sewers and unventilated cesspools, though mostly capable of destroying life under favorable conditions, are not, I think, responsible for much, if any, of the fatal effects properly attributable to sewer gas. Probably it is those constituents which analysis cannot find, and of which we know practically nothing, which impart to sewer gas its fatal capacity for bearing sickness and death to thousands of unconscious victims annually. This is an opinion, simply; let us see whether it will bear the test of examination.

<small>Chemical composition of sewer gas.</small>

Carbonic acid is the gas usually found present in greatest volume in sewers, both ventilated and unventilated. The proportion, as determined by analysis, varies according to circumstances, but it is usually large. This gas is an invariable product of the decomposition of all substances containing carbon. Its properties are so well known that I need give but little space to its description. Inhaled in concentrated form, it quickly produces death, and even when considerably diluted with atmospheric air, it produces asphyxia, and, unless the victim is quickly rescued from its influence, death follows promptly.

<small>Carbonic acid.</small>

This gas is the fatal "choke damp" of the coal mines, and deaths caused by it, in one way or another, are matters of almost daily occurrence. It does not readily leave sewers and cesspools, however, owing to the fact that its specific gravity is considerably greater than that of air, and so much of it as would naturally find its way into a house from a sewer, unless drawn in by a strong current of air, would not, probably, do much damage. At all events, carbonic acid is incapable of giving rise to the ordinary phenomena of sewer-gas poisoning.

The presence of an excess of nitrogen in sewers is readily accounted for by the fact that the union of atmospheric oxygen with the carbon of organic matter, forming carbonic acid, leaves it free. It is incapable of supporting animal life, but is not known to possess any poisonous properties. Nitrogen.

Sulphureted hydrogen, also a product of decomposition, is undoubtedly a very poisonous gas. Various experiments made with it have shown unmistakably its power to destroy animal life. One part in 250 of atmospheric air will kill a horse, and life may be destroyed by the absorption of this gas into the system through the skin pores, even though the lungs be abundantly supplied with pure air. But experience has also shown that even this deadly gas cannot be held accountable for sewer-gas poisoning. In laboratory work it is often necessary to make sulphureted hydrogen in large volume, and when the management of the apparatus is entrusted to students or beginners, the air becomes so strongly impregnated with its disgusting odor that one unaccustomed to the smell could not breathe it without serious discomfort. Indeed, a laboratory would not smell natural without it; and yet chemists, who breathe this and many other equally dangerous compound gases almost constantly while at work, have not been found to suffer any more from typhoid and gastric fevers, cholera, diarrhœa, general debility and other diseases known to be propagated by sewer gas, than those who never enter a laboratory. I have known instances in which students of analytical chemistry have been

Sulphureted hydrogen.

Chemists not especially subject to diseases caused by sewer gas.

made sick by inhaling sulphureted hydrogen, but not seriously, nor was their sickness of a kind similar to that produced by sewer-gas poisoning; and yet a house in which the smell of this gas was as strong as it usually is in many laboratories at any hour of day or night, would be considered untenable.

Nor can we charge the fatality of sewer gas upon the ammoniacal compounds which result from the evaporation, as well as the decomposition, of sewage. We must, then, seek for this most subtle and dangerous foe to health of all the gaseous emanations from the sewers, in what is called organic vapor. This is an indefinite name for something of which we yet know but little. Eliminate from sewer gas the organic germs which float in it, insensible to sight, touch and smell, and I doubt not it would be drawing the serpent's fangs. This vapor, so called, is doubly dangerous from the fact that we cannot tell exactly what it is. We can tell the exact amount of organic matter present in a gallon of sewage, but living organisms in sewer gas elude our senses and defy all but the most subtle and searching methods of analysis.

Ammoniacal compounds.

Organic vapor.

This brings us to a consideration of what is generally known as the germ theory of disease, which in this connection will be found to possess both interest and importance. For a full and complete discussion of this theory, the reader is referred to the very able treatise on "The Germ Theory of Disease and its Relations to Hygiene," read by Prof. F. A. P. Barnard before the American Public Health Association, and published in the report of that association for 1873.

The germ theory of disease.

For more than two centuries men of science have been steadily drawing nearer to the complete acceptance of the germ theory of disease. Many other theories have been advanced and discussed in the mean time, and some of them have been regarded as satisfactorily accounting for the origin and propagation of disease, but none have stood the test of the rigid scrutiny to which the close reasoners of the scientific world subject all theories and hypotheses. Some of them contained a measure of what we now regard as truth; others were extravagant imaginings, having no substantial foundation. At last the controversy narrowed down to a close and scientific comparison of the evidence in support of the chemical theory, of which Baron von Liebig was the most intelligent exponent, and the germ theory, originally advanced by Father Kircher, in his *Scrutinium Physico-Medicum contagiosæ luis quæ pestis dicitur*, and reduced to a scientific basis by Pasteur, the eminent contemporary and, on many points, the able opponent of Liebig; and the latter theory has gradually met with general acceptance. It is obviously impossible, as well as unnecessary, to follow this controversy and weigh all the evidence brought forward to support the rival theories, and I will merely outline what I understand to be the germ theory as now generally accepted. It presumes that disease is propagated by the invasion of the human system of algoid or fungoid forms, of microscopic proportions but possessing the power of rapid multiplication. The spores which proceed from these fungi, or the cells of the algæ, are carried by the air currents as the invisible pollen of flowers is carried, and, penetrating the human system, generate diseases. The fact that all forms of cryptogamic vegetation are propagated in this manner, may be regarded as, *prima facie*, favorable to the germ theory. Further evidence of the same kind is found in the results of Dr. Tyndall's experiments in transmitting the beams of the electric light through air and vacuo, by which he has shown that the former is charged with organic particles. Evidence of this sort is abundant and, as the rule, satisfactory, if not conclusive. Of proof we have not as yet enough to establish the germ theory as a demonstrated truth, but there are many facts which, it seems to me, can only be explained reasonably and rationally on this hypothesis. Certain diseases are known to be propagated by organic germs; in other cases it is probable, but not certain; in still others it is uncertain, if not doubtful; but we may, I think, accept with confidence the fact that a great many, if not all, diseases are communicated by living organisms which, in systems predisposed to disease or in a condition favorable to the development of disease, rapidly multiply, and, whether directly causing disease or not, are the media of its transmission and the vehicles of infection.

<small>Inorganic poisons incapable of producing zymotic diseases.</small> Probably the strongest of the many arguments in favor of the germ theory of disease is found in the fact that, in the whole range of inorganic substances, chemical analysis has discovered nothing capable of producing results in the human system in any degree comparable with those produced by the agencies which convey infection and produce disease. The action of the <small>Their action.</small> inorganic poisons is generally well known and definite. They destroy life or produce certain characteristic symptoms of derangement in the human system, but they are incapable of producing any of the diseases known to result from impurities imparted to air and water by the decay of organic matter. It may be claimed that the negative results of chemical investiga- <small>Proof and disproof.</small> tion prove nothing, but the most determined opponent of the germ theory of disease has never been able to produce, discover or describe any inorganic substance, elementary or compound, which could produce any one of the diseases attributed, and even directly traceable, to organic poisons.

Since writing the above my attention has been called to a <small>The gases of decay.</small> paper on "The Gases of Decay in some of their Sanitary Relations," read before the American Public Health Association, in October, 1876, by Prof. William H. Brewer, of the Sheffield Scientific School, New Haven, Conn. This paper is so clear and concise in presentation of the subject discussed in this chapter, that I am glad to be able to quote it in support of the views I have expressed. After discussing the composition of sewer gas, as determined by analysis, and showing that none of the gases yet described are capable of producing the phenomena of sewer-gas poisoning, Prof. Brewer says:

<small>Chemical action of sewer gas.</small> "If the physiological effects which follow the breathing of sewer gas, so called, are produced by actual gases acting chemically, then these gases are as yet absolutely unknown to chem-

ists, and if they exist at all, they are in too small quantities to be estimated by any known process of gas analysis. This, however, is no proof that they do not exist. The sense of smell tells us that there are organic gases and compounds never yet isolated, and of whose composition and properties other than their smell we are entirely ignorant. Indeed we are ignorant of the composition of most of the *smells* of putrescent matter. Smells. In the investigation of the gases from rotting fish, of which I have spoken, the gases were very stinking, intensely so, yet the actual amount of the gas which had the odor was too small to be detected by the ordinary means of gas analysis, and these analyses were conducted under the eye, and some of them with the aid of Prof. von Bunsen, then, as now, the most eminent gas analyst in the world. The analyses of sewer gases point in Analysis of sewer gas. the same direction. For example, the results of some experiments on the air of sewers and drains are given in the Report of the British Association Sewage Committee, 1869-70. Specimens were collected from various street and house sewers, chiefly in the Paddington district, and during August, so that there is every probability of the air being as foul as possible. They were chemically examined by Dr. W. J. Russell. The most impure air contained half a per cent. of carbonic acid; the remainder was oxygen and nitrogen, so far as discovered by analysis. Another 'with a foul smell' contained only one-eighth of one per cent. of carbonic acid. There were 'no combustible gases.' In their investigations they found only small traces of ammonia, and often no sulphureted hydrogen. It is needless to multiply cases. It is not, of course, denied that sewer gases have been found so concentrated and foul as to produce suffocation, Suffocation. but very bad effects are well known to often follow the admission of such minute quantities into our houses that they can barely be perceived, much less suffocate. That it lowers the tone of health and sometimes produces active disease in those who are subjected to it, is too well proven to admit of a doubt. So far as this first effect occurs (lowering the tone of health) we can easily imagine it to be produced by *chemical* causes. Definite physiological results are known to follow the absorption into the system of definite chemical compounds. The effect of medicines and of poisons are illustrations too common to need more than a reference to them. The agent may work speedily, as in the case of active poisons, or slowly, as in the case of cumulative ones. The effects may be gentle, as with certain tonics, or violent, and, as in arsenic poisoning, take a somewhat definite time, like a fever running its course; but in all poisoning by chemical means, the physiological effect is very largely proportional to the amount of the chemical used, and the effects cease with the victim. Moreover, the results are reasonably uniform.

<small>Poisons.</small>

<small>Effects of sewer gas on the human system.</small> "This is very unlike the effects believed to be caused by sewer gas or other 'filth gases,' where the results are by no means uniform, nor do they appear to be at all proportionate to the amount of the gas breathed, nor to its degree of concentration. More than this, the results do not stop with the victim; typhoid fever, once started, may extend to we know not how many other victims if the right conditions exist to carry it, and this brings us face to face with that mooted subject, the *germ* theory of zymotic diseases, a theory so generally accepted by chemists, so strongly combated by some of the most eminent microscopists and physiologists.

<small>Typhoid fever in Croydon.</small> "That typhoid fever has been caused by the escape of gases from sewers and cess-pools into houses, seems to me to be proven beyond a reasonable doubt. For illustration, in the now famous town of Croydon special cases are mentioned (ninth Rep. Med. Officer of the Privy Council, 104) where the disease is supposed to have been distinctly traced to this cause. The gas was known to have been driven into the house; it 'did not smell offensively, only a faint, sickly odor being recognized.' In this case the gas was driven into the house by a shower filling the conductors with water. Other cases at the same time are believed to be traceable to the same source. The odor was gen-

erally not rank, 'a faint odor alone being recognized.' I think it is generally conceded that typhoid, once started, may be propagated from patient to patient through the medium of the evacuations. Now all this is unlike the operation of any known chemical compound, gaseous or otherwise. Again (from the same report), the outbreak of cholera in the city of London, Union Workhouse, in 1866, investigated by Mr. Radcliffe, was shown to have taken place, in all probability, from a sudden efflux of 'sewer air from a drain containing choleraic evacuations,' this efflux being caused, or at least favored, by a sudden change of atmospheric temperature and pressure. Here again the gas, or 'sewer air,' spoken of as the agent, is not necessarily a 'gas of decay;' yet, if a gas at all, it must have been an organic gas, acting as a poison, but how unlike all actual chemical poisons, where the agent is a known chemical compound. *Cholera in London.*

"Again, decay of filth in the dark, and away from free access of air, is supposed to be productive of gases especially dangerous to health, more so than when the decay goes on in the light and free air; and, moreover, that sewer gas is rendered less hurtful by a free circulation of air in the sewers. That this last is not due to mere dilution, is shown by the deleterious character of the gas when diluted only after it enters the houses. *Organic decay in the dark.*

"Considered purely as a chemical question, these facts, if facts, are entirely inexplicable. If the germ theory is accepted, a plausible explanation is more easy. It is possible to imagine a condition of things in decaying organic gases similar to that which occurs in decaying organic infusions. It is known that such infusions soon swarm with minute organisms, the almost universal occurrence of which in such connection gave them the general name of "*infusoria*," and that different forms are generated according to the different chemical characters of the solution. The changing organic compounds in the fluid are doubtless the food with which these low organisms are nourished. Certain specific forms thrive best in certain definite *Chemical aspects of the question.*

Infusoria.

infusions, and appear there when given the proper temperature, and, once started, they increase and multiply as do other organisms. Now it is easy to imagine an analagous state of affairs in decaying organic gases. Moisture is always an element in the unwholesome gases of decay, and along with it are some gases that are organic, generated by the breaking up of the more complex molecules. Their quantity may be small compared with the whole volume of gas with which they are mixed, and yet sufficient to nourish floating organisms, just as a mere trace of solid matter dissolved in much water, making a very weak infusion, is often nutritious enough to support its swarms of infusoria. If this be the case, it may possibly explain the anomaly that dilution of gas with air within the sewer renders it comparatively harmless, while it may be very poisonous if it is diluted only after it enters our houses. Thus if the analogy is good that floating organisms, which may be the germs of disease, feed on and multiply in the decaying organic gases of sewers, as infusoria feed on and multiply in infusions when the temperature and degree of concentration are favorable, then such floating organisms, after having been once produced in the sewer, and then admitted into the house, would not be destroyed by dilution of the gases in which they float, while, on the other hand, proper dilution with air within the sewer might, by oxidation or in other ways, prevent their generation, or at least so impair the conditions that they cannot multiply in harmful numbers.

"The belief that malaria is related in some way to the gases of decay, has already been referred to. That it is often so associated in moist air is well enough known. The draining of swamps and giving the air access to the vegetable mud accumulated in such places, the clearing of land and consequent rapid decay of the accumulated leaf-mold, have often been related to the existence or spread of malarial diseases. Even the decaying leaves of our shade trees in the streets are often accused of adding to the malaria of a region. In these cases the decay goes on in free air and light, and the gases are diluted to the last degree as soon as liberated from the generating mass. Yet here, too, we can understand how organic gases may be concentrated enough, before being poured forth into the atmosphere, to give the requisite nourishment to the organisms or "germs." Such decaying vegetable matter is very porous; it contains air as a sponge may water, and this air, permeating the decaying substance, cannot be otherwise than highly charged with the products of decay, ready to be driven out in several ways. Take rotten wood as an example: the measure of its porosity is seen in the difference of weight when wet and dry. A little experiment tried for another purpose the present week may be used as illustration. A few days ago, where some workmen were repairing the wooden pavement in one of the streets of our city, I picked up a few pieces of the half-rotten wooden blocks. They were saturated with the water of the recent rains. Two pieces weighed, as they came from the pavement, respectively 287 and 130½ grams. They were then left on the table in my study four days, and then yesterday weighed again. They were not yet dry by any means, yet they weighed respectively only 154 and 54½ grams. That is, as thus dried, one will absorb 86 per cent., the other (and most decayed) 139 per cent. water, before saturation. It is easy to see how much foul air in a concentrated form a half-rotten wooden pavement may hold, to be driven out by the first shower or by any other cause that disturbs the equilibrium of the atmosphere." *Decaying vegetable matter.* *Wood pavements.*

The deadly outbreak of typhoid fever at Over-Darwen, England, in the early part of 1875, gave rise to a discussion of much interest on the origin and propagation of febrile diseases. Professor Tyndall, with characteristic enthusiasm, opened the controversy by announcing it as his opinion that the organisms which are assumed to be the immediate cause of fevers, are evolved from living blood corpuscles by a process of development, and that such diseases have their origin in the human system, practically independent of surrounding miasmata or *Typhoid at Over-Darwen.* *Tyndall's theory of disease.*

emanations from decomposing organic matter. It soon became evident, however, that the weight of scientific opinion was on the other side. Dr. Lionel S. Beale, whose authority in such matters is much higher than Professor Tyndall's or perhaps any other scientific man in Great Britain, seems to have taken a middle ground, and to have discovered that the practical are of more importance than the scientific aspects of the question. That the poison of fever grows and multiplies after its kind like other living things is, he says, a fact established beyond controversy; although it is still a matter of dispute among original investigators whether it is a microscopic fungus originating without, or a living particle rising within and from the living matter of, the human system. But whatever may be the truth on this point, he insists that human beings are alone responsible for the production of these germs and for their maintenance and spread; and that a state of society is conceivable in which fever germs would neither arise nor multiply, or, in the event of their introduction *ab extra*, would themselves perish instead of damaging or destroying the higher life.

"Fever germs," he says, "will not be developed direct from filth, but by permitting people to live year after year in open defiance of well-known sanitary laws, the generation of fever poison in their bodies is favored, while its free growth and multiplication if imported is reduced to a certainty. It is therefore our aim to prevent people from falling into that condition of health which favors the organization and propagation of contagious fever poisons in their bodies. * * * Although we may successfully and without fear contend with fever germs if we only preserve our healthy powers of resistance, hundreds of human organisms are, through defective sanitary arrangements, being prepared for invasion." "Bad air and sewage, the adjacent dung-heap," concludes Dr. Beale, "may all be perfectly free from fever germs, but nevertheless certainly will bring about changes which will render many of those exposed to their influence the ready victims of disease. While, therefore, it is desirable, by the use of disinfectants and by other means, to destroy existing fever germs with all possible speed, it is certainly of far higher importance, as regards the welfare of the people, that we should do our utmost to press upon authorities the necessity of providing pure water and efficient drainage wherever men congregate. Good water and well-arranged sewers render impossible such a calamity as that which we have now to deplore at Over-Darwen. Even though the inhabitants of a town well drained and supplied with good water should be fully exposed to the assaults of hosts of fever germs in their highest state of morbid activity, they would suffer no injury."

The diseases commonly supposed by the medical profession to be caused or propagated by the organic germs resulting from decomposition, and which we may assume are always present in sewer gas, are chiefly those classed under the generic name of zymotic diseases—a name derived from a Greek word meaning to ferment. A zymotic disease is any epidemic, endemic, contagious or sporadic affection produced by some morbific principle acting on the organism and producing results similar to fermentation. The most careful investigations and experiments, extending through many years, have led to the conclusion, generally accepted by the medical profession in all countries, that gaseous poisons (*venena aërea*) give rise to many diseases of the zymotic class, among which are the following: *[margin: Zymotic diseases.]*

Cholera Asiatica,	Scarlatina,
Cholera morbus,	Diphtheria,
Cholera infantum,	Measles,
Dysentery,	Typhoid fever,
Diarrhœa,	Typhus fever,
Small-pox,	Yellow fever.

[margin: Diseases communicated by sewer gas.]

These are by no means the only diseases of the zymotic class which might be placed in the list, nor are those given arranged in the order of their relative importance. In times of epidemic,

Asiatic cholera is the most fatal, while the milder forms of cholera are more frequent and, in the aggregate, probably more destructive to life. At the time of this writing small-pox and diphtheria are raging as fatal epidemics in many parts of the country, especially in and about New York and in the neighboring cities of New Jersey and Long Island. There is no doubt that the prevalence of these diseases is attributable, primarily, to defective drainage. Many parts of the country are yearly visited by yellow fever, and the mortality of this and kindred diseases is greatest where the most unhealthy conditions prevail. Nearly all the diseases on the list have at times assumed the form of epidemics. Typhus and typhoid fevers are not known to be contagious, but they are among the most frequent results of sewer-gas poisoning.

Prevailing epidemics.

Those who accept the germ theory of disease, even in part, can find in it a satisfactory explanation of the deadly power of sewer gas. Dr. Thomson, a careful experimenter, found organized forms in the air of the sewers of London, and although this proves nothing in itself, it is significant when taken in connection with the fact that, during the prevalence of epidemic diseases in England, their propagation has been carried on most rapidly through sewers and house drains. A similar phenomenon occurred under the observation of the writer a few years ago, which may serve as an illustration of the transmission of infection by means of sewer gas.

Germ theory explains the dangers of sewer gas.

On the outskirts of a beautiful and generally healthy town near New York there was a long row of tenements under one roof, divided into thirteen separate dwellings. These dwellings were occupied by the families of men employed on the railroad and about the extensive coal docks not far distant. In the spring of 1872 a case of typhoid fever made its appearance in one of the families; soon another, and then several were reported, and the number increased until every house in the row was affected. An examination of the premises was then made by a committee of local physicians, at the request of the

Typhoid fever communicated through a house drain.

town authorities, and it was found that under the houses was a brick drain common to them all. Connection between this drain and the waste pipes of the houses was established by means of glazed earthen pipe of small size, in which the joints had been made with bricklayers' mortar. None of these joints were tight. In each house there was one slop hopper, which also discharged into the drain. The plumbing work had been done in the cheapest manner possible, and while the waste pipes leading from the kitchen sinks and slop hoppers of each house were trapped, the pipes were so thin and had sagged so much by their own weight that in every case but one a dash of water would have been quite likely to empty the traps, either by direct downward flow or by syphoning. Communication was thus opened from one house to another through an unventilated drain, and although intercourse was not permitted between the inmates of houses free from fever and those in which it had made its appearance; the germs of disease, contained in the excreta of those first attacked—which had been thrown into the slop hoppers—had found a means of access which had escaped the vigilance of the physicians.

In country districts the germs of typhoid fever are most frequently communicated to the human system through drinking water drawn from wells and springs contaminated with organic impurities. This also occurs to a great extent in cities and small towns not provided with water works, and in which the inhabitants are dependent upon wells necessarily sunk in close proximity to privy vaults and cesspools for the reception of house drainage. In cities drained by sewers and supplied by water works there is little chance, except when the most inexcusable carelessness has been manifested by builders and plumbers, of any contamination of the water, unless drawn from polluted sources, and if the sewers are well ventilated and the house connections properly made, there seems to be no reason why the public health should suffer from any cause traceable to sewage. But sewers are seldom well ventilated in

Disease germs in drinking water.

Sewer gas poisoning preventable in cities.

this country, and house connections with sewers are not always, if often, so made as to offer effective opposition to the inflow of poisonous gases generated in these channels of pollution under our streets. The first step in the direction of reform should be the thorough ventilation of the sewers, and then any improvement in the character of the plumbing work in houses and dwellings will be more likely to accomplish the results desired.

The drainage of Croydon. The history of the sanitary works of Croydon, England (before referred to in an extract from Prof. Brewer's paper), shows very clearly the importance of sewer ventilation. This was formerly a very healthy town, but as the work of providing it with sewers approached completion an epidemic of typhoid fever broke out. Latham, in his excellent work on Sanitary Engineering, which I consider one of the most valuable of recent contributions to the literature of this very important subject, gives the following account of the effect of ventilating the sewers in diminishing the death rate of that city:

"The mortality of Croydon rose from 18·53 per thousand in 1851 to 28·57 per thousand in 1853. Those early sewer works were designed on the principle that all matters were to be so *Formation of sewer gas not prevented by flushing* rapidly discharged from the sewers, and the sewers flushed with such a copious supply of water, that decomposition could not take place, and therefore it was thought that sewer gas would never be present; but in practice this theory was not found to be borne out, and it is a remarkable coincidence as to the cause of the frequent outbreaks of fever in Croydon which took place at certain intervals until the year 1866, when the sewers were thoroughly ventilated, that diseases which formerly made their haunt in the low-lying districts were transferred, *Typhoid fever following the lines of sewers.* after the completion of the drainage works, to the highest or best portions of the town, thereby establishing the fact that the presence of the disease in the high localities was due to something carried in the air of the sewers, which, in obedience to a natural law, accumulated in the highest part of the district.

It may be said that, as Croydon was sewered on the small-pipe system, the result of non-ventilation was attended with more marked results than is the case in towns where sewers of larger size are in vogue, as the fluctuations in the rate of flow and the effects of sudden changes of temperature, which have an extraordinary influence on the air of sewers, in this case exercised a more marked effect in increasing the pressure of the imprisoned sewer air.

"With regard to the results that have arisen when ventilation of sewers has been adopted, the case of Croydon shows clearly that proper ventilation has been attended with very beneficial results. Since the introduction of systematic ventilation there have been no periodical outbreaks of fever, and the general rate of mortality has so declined that, in a district having a population of nearly 60,000 persons, the rate of mortality rarely exceeds 18 in the 1000, which is a standard of health unparalleled in sanitary science for a district having so large a population. The example of London affords another striking example as to the influence of sewer ventilation. Here the sewers are ventilated, though no general plan is adopted for dealing with the noxious effluvia escaping from the ventilators, and yet London stands at the head of all large towns by reason of its small death rate, which has been ascribed by more than one eminent authority to the somewhat rude ventilation provided for the sewers." *Beneficial effects of sewer ventilation.* *The example of London.*

I think I may say, without doing any injustice to our civil engineers, that sewer ventilation is not appreciated in this country in proportion to its importance. Certain it is that our public authorities are too generally indifferent to it, and no plan has been proposed which has secured general adoption, except that of ventilating through perforations in the man-hole covers. Most of those who have ventured to propose plans of sewer ventilation have been ignorant of the conditions which must be observed to secure the results desired. Like most matters connected with sanitary engineering, the ventila- *Sewer ventilation not appreciated in the United States.*

tion of sewers seems to be more generally discussed than understood. Every little while men who should know better write letters to the newspapers, urging upon the attention of the health authorities the plan of ventilating sewers through the chimneys of furnaces provided for the purpose, and illustrating the practicability of the system by comparing sewer ventilation to the ventilation of mines. This oft-invented plan is an old one, having been practically tried and abandoned several years ago in England. The fallacy of the idea is found in the fact that the ventilation of a system of sewers presents us a problem in no respect analogous to the ventilation of mines. In a mine we can direct an air current from passage to passage along its entire length, so that each gallery shall be swept over by a current of air entering at the downcast shaft and drawn out at the upcast. A system of sewers, on the contrary, may be likened to a tree and its branches. From the main sewers extend the laterals in all directions, and these, in turn, become the mains for smaller sewers, until a whole district is covered with an intricate network of sewers and drains, ending, in whatever direction we may follow it, in the soil pipe of a house. These laterals vastly exceed, in their aggregate area, that of the mains into which they discharge. Now, to produce even a perceptible flow of air through these laterals in the direction of ventilating shafts established at points on the main sewers—say, a current moving at the rate of one mile an hour—we should have to maintain in the main sewers a perfect hurricane. Any engineer who will study a plan of the sewage system of any of the New York wards, will see at once the hopelessness of an attempt to apply mine ventilation to them.

Another and very important objection to this plan is found in the fact that even if we could maintain by artificial means an outflow from the sewers, we should not secure as good a result as would be attained by simply providing openings and allowing the sewers to ventilate themselves. The sewers are

Impracticable plans for sewer ventilation.

Objections to exhaust chimneys.

continually "breathing." At times they draw air in with considerable force; at other times they expel the air with a force great enough, if resisted by obstructions at the openings in the man-hole covers, to displace the seal of any water trap in use. Hence the folly of seeking to produce by artificial means effects which nature is ready to accomplish unaided, if we will only give her an opportunity. Furnaces, blowers, exhaust fans, and other mechanical apparatus for exhausting the air in sewers or forcing fresh air into them, are expedients which suggest themselves to those ignorant of the construction of sewers and of the operations going on within them. They have all been tried with extreme care and have failed utterly, and yet sewers may be perfectly ventilated by means of simple openings. If every man-hole cover in our streets were replaced with a grating, as open as might be consistent with strength, and these were kept free from ice, snow and mud by men employed for the purpose, and every house owner were required to vent his soil pipe (unobstructed by any form of trap along its line) above his roof, the ventilation of our sewers would be accomplished. The labor spent in devising other means of attaining this result is very likely to be wasted.

The only possible objection to sewer ventilation is, of course, based upon the assumption that the air from the sewers would carry with it noxious gases and organic germs to poison the surrounding atmosphere. This objection is less serious than is commonly supposed, as the organic germs in sewer gas are quickly oxidized and destroyed by contact with fresh air.

The question of sewer ventilation is, however, one which the individual citizen is not called upon to decide. All that he can do in the matter is to see that his house is properly drained; and in such portions of this book as relate to drainage, the author has endeavored to deal with existing conditions—one of which is assumed to be defective sewer ventilation.

Were it desirable to multiply testimony respecting the dangerous effects of sewer gas when permitted to mingle with the confined air of our—as the rule—badly-ventilated dwellings, this chapter might be extended to indefinite limits. The fact is too generally conceded, however, to need the confirmation of abundant proof.

In determining whether sewer gas is present in the air or not, dependence is too generally placed upon the sense of smell.

<small>The absence of foul smells no proof of good plumbing.</small> The nose is a much-abused but very useful organ, and, if properly cultivated, is of great service in the work of sanitary inspection; but the fact that the air of a house is ordinarily free from unpleasant odors does not prove that disease germs, generated in the sewers, cannot find means of access to them. Two instances will serve by way of illustration. In 1874 a manufacturer, doing business in one of the upper wards of New York, had occasion to employ crude petroleum for some purpose, a portion of which, being useless to him and of little <small>Petroleum in a New York sewer.</small> intrinsic value, he allowed to run into the sewers. Immediately complaint was made to the Sanitary Superintendent of the Board of Health that every house in the neighborhood was filled with the smell of petroleum residuum, and petitions were sent in praying that the nuisance might be abated and the manufacturer enjoined from running any more of the offensive refuse into the sewers. The smell of the petroleum could only find its way into the houses through waste pipes communicating <small>A nuisance reveals a danger.</small> directly and indirectly with the sewers, and its presence only called attention to the fact that the traps afforded no effectual protection against the influx of sewer air. The gentleman at that time filling the office of Sanitary Superintendent, suggested to the committee of house owners and tenants who waited upon him with their petitions, that the offending manufacturer deserved a vote of thanks for having called their attention to an evil of which they had previously been ignorant; but they insisted upon having the nuisance abated, and when it was done they probably relapsed into the comfortable indifference to the condition of the plumbing work in their houses from which they had been temporarily aroused by an unpleasant but, in itself considered, harmless smell.

The other case is somewhat similar to the one already mentioned. At the time of this writing there is an oil refinery on the Long Island side of the East River, which runs the unsalable residuum of the refining process into a sewer which drains a populous section of the Eastern District of Brooklyn. When the tide is low, uncovering the mouth of the sewer, and the wind is from the west, the smell of petroleum is unpleasantly noticeable in the bath-rooms and water-closets of nine-tenths of the houses in a district of considerable extent and containing a large resident population. Wherever the smell makes it way, we may safely conclude that it carries with it poisons capable of producing sickness and, under favorable circumstances, of destroying life.

Petroleum in a Williams burgh sewer.

In the discussion of questions connected with sanitary engineering and public health, it is important to avoid a confusion of ideas growing out of a failure to discriminate between the abuses of the sewage system and the system itself. We should not be led by failures in engineering and plumbing to denounce the modern improvements which have rendered water service and house drainage possible. Breslau and Munich are to-day suffering nearly double the death rate of London and Vienna; and yet Breslau and Munich have but little house plumbing, and place their cloacal nuisances outside their dwellings. The population of London, on the other hand, depends almost wholly upon indoor systems of water-closets and house drains; while Vienna has reduced its mortality over 30 per cent. by the provision of an over-abundant water supply from the mountains, flushing the house drains and sewers and supplying pure water in excess of all wants. London, Paris, and the parts of Glasgow, Berlin and Liverpool which have an ample water supply at high pressures, are for the same reason as Vienna nearly as healthful as the most favored rural districts. Here also we find the best sanitary engineering and the best plumbing.

Conditions affecting the health of cities.

CHAPTER III.

WASTE AND SOIL PIPES.

Importance of good drainage. Those who have given the subject of house drainage any attention do not need to be told that while the plumber may unskillfully perform many parts of his work with no greater resulting mischief than a perpetual inconvenience to those who live in the houses in which he makes his ignorance conspicuous, any deviation from good workmanship in the waste pipe system may prove a perpetual menace to health and even life. In the *Traps and seals.* succeeding chapter I shall speak of traps and seals, and of the method of ventilation by which the gaseous emanations of the sewers are conducted away from the points at which they are otherwise likely to exert a pressure great enough to displace or pass through the water seals upon which we rely, to a great extent, for protection against the inflow of foul air currents. In this chapter I shall speak of the selection of suitable pipes, the uniting of the lengths by water and gas tight joints, and the avoidance of all causes of obstruction and leakage.

Lead waste pipes. For small waste pipes the material generally employed is lead. It can be easiest bent, joined, cut and otherwise manipulated, and cannot do any damage by parting with poisonous salts to the waste water brought in contact with it. The size and weight of lead pipe used should be determined with reference to the service expected of it. A waste pipe should never *Relation of size and weight to volume of discharge.* be so small as to retard the outflow of water or become easily choked; nor so light as to be rapidly destroyed by corrosion, or broken from any cause. In my judgment, based upon a somewhat careful examination of good plumbing work, the best sizes and weights for general use are the following:

Sizes adapted to various uses.

	Diameter.	Weight per foot.
For bath wastes	$1\frac{1}{2}$ in.	3 lbs.
" " overflows	$1\frac{1}{4}$ in	$2\frac{1}{2}$ lbs.

WASTE AND SOIL PIPES.

	Diameter.	Weight per foot.
For basin wastes	$1\frac{1}{4}$ to $1\frac{1}{2}$ in.	$2\frac{1}{2}$ to 3 lbs.
" " overflows	$1\frac{1}{4}$ in.	$2\frac{1}{2}$ lbs.
" wash-tub wastes	2 in.	3 lbs.

For kitchen sink wastes it is generally advisable to use three-inch pipe, with the exception of the section containing the trap, which should be two and one-half inches in diameter, in order that anything forced out of the trap shall not lodge further along and again obstruct the flow in a less accessible place. It is a very common practice among slovenly kitchen servants to remove the brass strainers and brush all sorts of greasy waste matter not valuable as soap grease into the pipe. Consequently its stoppage by an accumulation of solid matter in the trap, which defies boiling water and *sal soda*, is in many houses a matter of almost daily occurrence. These accumulations will sometimes form even with good and careful management, and to avoid serious trouble it is desirable to make the trap at least one-half inch smaller than the body of the pipe below it, and place it near the strainer so that a short flexible rod of any kind may be passed into it. When trap screws are provided this precaution is unnecessary, as accumulations can then be cleaned out of the trap without forcing them into the pipe. A still better arrangement is what is known as the grease trap, which has many advantages but is not generally used. This is a vessel receiving the waste water of a sink and having its outlet from the bottom through a pipe rising nearly as high as the top of the trap. As greasy substances are lighter than water, they float and assume a semi-solid consistence inside the receiving vessel while the water below passes off readily. This device saves the fatty matter, which for soap making is of considerable value in the domestic economy.

Traps in kitchen sink wastes.

Obstructions in sink wastes.

Trap screws.

Grease traps.

For large waste and soil pipes, cast iron is the material which for about fifteen years has been almost exclusively used in New York and vicinity. Its advantages over lead are many. In the first place it is cheaper, which is an important consideration;

Cast iron waste pipes.

WASTE AND SOIL PIPES.

but its chief superiority consists in the fact that it is lighter, stiffer, stronger and less liable to accidental injury. Thin lead pipe of large diameter is necessarily weak. It sags by its own weight and often breaks from inherent weakness. Heavy lead pipe of sufficient size is both too costly and too heavy for use, and, owing to the weakness of the material, there is great difficulty in securely fixing it in any position in which its weight falls upon the fastenings. Iron pipe is wholly free from these objections. It is strong, stiff, and corrodes so slowly that, if of proper weight and good quality, it will ordinarily last as long as a house. We could scarcely wish for a better material. It is easily cast into any desired shape, and the requirements of plumbers have been so fully anticipated by the founders that it is very rare to find an architect's specifications calling for any form of pipe, joint, trap, Y, branch or bend, which is not kept in stock, or which, if not in stock, cannot be fashioned by combinations of parts already made. One of our iron founders told me, only a few days previous to this writing, that he was almost daily in receipt of orders to cast special shapes of pipe sections and joints, but rarely found it necessary to make a new pattern. If he did not have exactly the article wanted, he could almost always make it by putting two or three pieces together, no matter what the use for which it was intended or the position in which it was to be placed.

Lead not adapted for large sizes.

Superiority of iron pipe.

Iron cast in every required shape.

But while we have in cast iron a material which, all things considered, is as good as we could want, the efficiency and safety of a line of cast-iron soil pipe depend in a great degree upon the manner in which it is set up and joined. Probably every practical plumber in the country knows how to make a tight and permanent joint in iron pipe, which, for convenience in handling, is cast in short lengths, but they do not always do it. In some instances I have seen the lengths set into each other without any attempt to make joints at all. The consequences of such slovenly workmanship may be serious and far reaching. Such connections, while they do not often leak water—which

Conditions of safety in the use of iron pipes.

Careless joining of lengths

WASTE AND SOIL PIPES.

is much to be regretted—afford a free outlet for the gases of the sewer and drain. These gases rise along the sides of the pipe, spread between floor timbers and practically distribute themselves over the entire house, vitiating and poisoning the air of living and sleeping rooms. There are many houses which are saturated, so to speak, with sewer gas escaping from loose or defective joints in soil pipes, and the evil is one against which architects and builders, as well as plumbers, should carefully guard. *Escape of sewer gas from defective joints.*

There are several ways of making good joints in cast-iron pipe. There are also several methods in common use which cannot be depended upon, and which should not be employed under any circumstances. Those which are good, as well as those which are not, may be briefly described as follows: *Methods of making joints.*

In making a joint with melted lead, a gasket of oakum, tow or other fiber should be inserted in the cavity of the hub, and the spigot end of the length next above it set firmly down upon it; or the gasket may be forced in with a suitable tool after the lengths are set up. Its utility is to keep the melted lead from running out of the joint and obstructing the bore of the pipe at some point below. Upon the gasket the lead is poured from a ladle, and at this point too many plumbers rest satisfied, thinking they have made a lead joint. Lead, however, is a metal which shrinks in cooling; moisture or dirt will prevent its adhering to the iron, and no good workman will consider the joint made until he has finished it by carefully calking all around it. With a suitable calking tool, the lead is expanded cold until it fills the joints as perfectly as a gold plug fills the cavity of a tooth upon which a skillful dentist has operated. *Lead joints. Gaskets. Shrinkage of lead in cooling. Calking.*

The tightness and strength of a lead joint depend in great degree upon the use of plenty of lead and the manner in which the calking is done. In good work it is necessary to use a pound of lead to each inch of the inside diameter of the pipe. Such, at least, is the general experience. A joint in 3-inch pipe requires three pounds of lead, in 4-inch four pounds, and so on *Quantity of lead required.*

up to the large sizes. In calculating the amount of metal to be poured, it is safe to allow one pound in three, or two pounds in six, for waste. The amount which a good workman will spill in pouring is not, probably, as great as this, but one pound in three is not an excessive margin for loss.

Allowance for waste.

In the opinion of many good workmen, a joint should never be filled by two pourings when it is possible to make it with one ladleful. This is a mistake. When a little extra time can be given to make the job first class, two pourings are better than one. I know a very expert practical plumber whose method of calking in first-class work is to pour in enough metal to form a ring, say half an inch thick, calk this all around, then fill the joint with metal and calk again. In iron pipe, a good deep hub with a small play between it and the ring on the spigot end, is the best for good work. When the pipe goes together loosely, there is more difficulty in keeping the lead from running into it and obstructing the water-way. There are several firms in this country having a high and well-deserved reputation for the cast-iron pipe made by them.

Filling joints with two pourings.

Good fitting pipe the best.

The use of alloys which expand in cooling has been suggested for making joints in cast-iron pipe. The idea is a good one theoretically, and would, doubtless, be found to work practically. The principal objection to this method is that most of the alloys which possess the property of expanding in cooling are expensive. Old type metal is probably the cheapest form in which such an alloy can be bought, but is too hard and brittle owing to the percentage of antimony it contains. Such joints would enable careless or slovenly plumbers to dispense with the labor of calking; but lead properly put in and expanded by tools of the right shape, is good enough and will last as long as the pipe.

Joints made with expanding alloys.

Joints well calked with red lead are tight and durable under ordinary conditions. Mixed with oil to the proper consistence, about that of fresh glaziers' putty, the red lead is worked down into the joint with a calking tool, and unless the space to be

Red lead joints.

filled is very large, the joint will be tight. I do not, however, consider this as good as the method previously described, for the reason that, if the pipes are moved a little, or shaken from any cause, there is great danger that red lead joints will be broken. Such possibilities of accidents must be guarded against.

Putty joints are sometimes met with in very cheap plumbing work; but as such joints should never be made, the method of making them need not be described. The same is true of mortar joints, which are, if anything, worse than no joints at all, because they give rise to a mistaken sense of security among those who are so ignorant and at the same time so credulous as to believe whatever a dishonest contractor may tell them. With the expansion and contraction of a pipe, mortar or cement cracks and leaves the joint practically open. I would condemn with equal severity joints made with Portland cement when lengths of cast iron pipe are to be united. This cement is probably the best substance for uniting lengths of earthen pipes for land drainage and other purposes, but it is not adapted to iron. It is pervious to both gases and liquids to some extent, and as it expands and contracts differently from iron, it soon cracks and crumbles, leaving the joint open. *Putty joints. Mortar joints. Cement joints.*

Joints made with rubber washers have been used to a considerable extent in the West, and attempts have been made to introduce the washers in the New York and other Eastern markets, but they have not been regarded with much favor. The washers are rings of vulcanized India rubber, cylindrical in section, so that they easily roll when slipped on the end of a pipe, and completely close the space between the spigot and the sides of the hub into which they are inserted. Being elastic, these rings permit any amount of contraction and expansion possible in cast-iron pipe, and probably keep the joint tight for a considerable time. The principal objection of intelligent plumbers to this form of joint, is based upon the fact that it is difficult to procure pure India rubber, well cured. How far this objection is valid I am not in a position to say. Were the rings *Rubber washer joints.*

intended for use in positions in which they would be constantly submerged, I should not hesitate to trust them. As it is, I have doubts. Lead is certainly a great deal safer and, consequently, a great deal better.

Rust joints. Rust joints may be made by means of a mixture of *sal ammoniac*, sulphur flowers and iron borings. Such joints are tight without doubt, since they make a line of pipe one continuous and rigid piece. It is for this reason that many plumbers object to them. They assert that a pipe united with rust joints **Difficulty of separation.** cannot be taken apart without great danger of destroying it, as the breaking of a well-made rust joint is an extremely difficult operation, and that if it is necessary to open the pipe at any point, the whole may be destroyed. Rust joints can be separated, but it is a delicate and difficult task, involving much care and labor, and one always attended with some danger to the pipe. This kind of joint is, I believe, the only one which can **Rust joints best in pipes which receive exhaust steam.** be depended upon in pipes which receive the exhaust of steam engines. In such cases lead joints are almost certain to come apart. With regard to the difficulty of separating them, it does not seem to me much greater than in the case of well-calked lead joints. The only way to get a lead joint apart is to melt it, and the pipe is often so placed that the plumber dare not apply fire to it. In that case it must be broken.

Sulphur and pitch joints. Sulphur and pitch joints have been made with a composition consisting of equal parts of these substances. This composition is used to some extent in the arts for making joints analogous to those in soil pipes, and if we must have cheap and bad work there is no doubt that sulphur and pitch joints would be **Cheap filling for joints.** far better than those made with putty, mortar or cement. Other cheap substances might be used, but when the plumber ventures any experiments in this direction he should be sure **Qualities required.** that the material he uses is not decomposed by sewer gas, that it contains no volatile constituents, and that in drying it does not shrink nor crack nor become rigid and unelastic. Until he has determined these qualities by tests which admit of no mis-

WASTE AND SOIL PIPES.

take, he would do well to use only the best materials employed by the trade and which experience has shown to be adapted to the purpose.

In setting up a line of soil pipe, intelligent provision should always be made for the expansion and contraction of the metal resulting from changes of temperature. These changes, however, are seldom sudden or extreme, but when the pipe is at any point rigidly fastened to the wall it expands in both directions. The amount of motion at the ends is small, but it must be provided for or it will provide for itself. The power with which iron expands as its temperature is raised, is practically irresistible. The end of a pipe may not move more than an eighth or a sixteenth of an inch, but the power with which it moves that distance is so great that it can only be resisted by a power great enough to crush the metal. This would be, in ordinary cases, equal to about 75,000 pounds per square inch, the strength of cast iron to resist crushing strains being from 60,000 to 90,000 pounds per square inch. Consequently, we see that unless the fastenings at the ends of a line of cast-iron pipe are of such a character as to admit of slight movement, something must give way, and it is not likely to be the pipe. This, then, must be provided for in the character and position of the fastenings, which must be so arranged that, while allowing for some movement, they shall not develop a tendency to break or loosen the joints. Under ordinary conditions the amount of expansion is seldom great enough to give much trouble, but when steam or a great volume of very hot water wastes into an iron pipe, it is sometimes great enough to loosen joints and even crack the pipe. *Expansion and contraction of iron pipe.* *Conditions favoring greatest expansion.*

In good practice a vertical line of soil pipe should be set on iron flanges, one to each length. There is always more danger of having too few than too many. In much of the work which has come under my notice, dependence has been placed almost wholly upon pipe-hooks, which are very good for keeping a pipe steady, but are not to be trusted for carrying any consider- *Supports for vertical lines of iron pipe.* *Pipe hooks.*

able weight. There are sundry patented devices for sustaining vertical pipes against flat walls, in recesses and in corners, which, so far as I can speak from experience, are mostly useful and practical inventions. Of the proper methods of supporting lines of pipe laid on grades nearly horizontal, I shall speak further on.

Patented devices for supporting pipes.

When the small waste pipes are of lead, and the main waste or soil pipe is of iron, as is found in most examples of good plumbing practice, it is necessary to make connections between them which shall be tight, strong and durable. This is a matter of great importance, but in much of the cheap contract work of the time we are very apt to find careless workmanship at this point. A safe method of connecting lead and iron pipe is by means of the tinned brass ferrule. The taper end of the ferrule is slipped into the end of the lead pipe and soldered fast. The other end, which is provided with a flange, drops into the hub of the soil-pipe branch and is secured in place by a well-calked lead joint. Copper and iron ferrules are also used; but while copper is better than iron it is not so good as brass. A joint made by this method is safe and durable, and will never need repairing until the pipes themselves give out. A cheaper method, and one which is probably just as good when properly employed, is to substitute an iron ring for the ferrule. In setting up a line of iron pipe, it is almost always necessary to cut off one or more ends, making rings a few inches wide. One of these rings, or collars, is slipped on the end of the lead pipe, which is turned over it. Thus reinforced, the end of the lead pipe is slipped into the hub of the soil-pipe branch, and the joint is calked in the ordinary manner—the iron collar answering every purpose of the ferrule, and utilizing an otherwise waste piece of metal. Perfect joints may be made by this method, which can be recommended with confidence.

Connecting iron and lead.

Brass ferrules

Copper and iron ferrules.

Ends of pipe as a substitute for ferrules.

Much of the objection which theoretical sanitarians have made against the use of cast-iron soil pipes is founded on the belief that, because they cannot be soldered together or to lead,

No objection to iron and lead connections.

WASTE AND SOIL PIPES.

it is impossible to make tight joints between the lengths or with lead connections. No such difficulty exists in practice. Every plumber knows this, and the statement is made merely because a great many who are not plumbers have mistaken ideas upon this subject.

In connecting an iron water-closet receiver or hopper with a lead waste containing a trap, we have a problem analogous to that already considered, but the method employed must be somewhat different. The pipe is brought up through the floor and, in first-class work, through the bottom of a lead safe, to which it is soldered. When no safe is used, the end of the pipe is expanded, so as to form a broad flange resting on the floor, and this is thickly covered with putty. The nozzle of the hopper or receiver is then set into the mouth of the pipe, and is firmly secured to the floor by means of screws which pass through the iron flange and through the flange of lead formed by expanding the pipe. If the putty is properly applied and the closet firmly screwed down, the joint will be tight and permanent. Good white-lead putty, tightly compressed, will make a joint which will be steam tight for years, and I know of no good reason why such a joint should not be both water and gas tight as long as it remains unbroken. India-rubber washers have sometimes been used in such joints, but I should consider putty, properly applied, quite as good, if not better. When lead safes are used, the lead pipe, as I have already said, is soldered fast to it, with or without flanging, as the plumber may prefer. The joint is then made as already described. In no case should the weight of the pipe, with its water seal, be allowed to come upon the joint. If not secured to the floor, it should have an independent support of some kind under the bend of the trap. Among the instances of bad plumbing which have come under my notice, I have seen cases in which the whole weight of a 4-inch lead pipe connecting a water-closet with a soil-pipe branch—so thin as to be scarce able to sustain its own weight and with four or five pounds of water in the

Connecting water-closets with waste pipes.

Putty.

Rubber washers.

Joints not made to sustain pipes.

Examples of bad plumbing practice.

trap—has been allowed to hang from a joint in which the only dependence was upon the cohesion of putty. That only a fool or a knave would do such work—even by contract—is a fact too evident to need the support of argument.

In the ordinary plumbing of a house there is no part of the work which is so liable to be slighted, or so likely to give rise to conditions unfavorable to health, as the overflows. In jobs of work of otherwise unexceptionable character, the overflows are often the weak points, and through them the sewer gas finds its way into the house. In the cheaper classes of work, both in New York and Brooklyn, I have seen houses so piped that the overflows had direct and untrapped communication with the sewer. When safe wastes are connected with traps, they are at times as great a source of trouble as the overflows. With the trap dry, from any cause, it may be necessary to put in the plug of the fixture to make a seal, which would be effectual but for the overflow and safe waste, which admit the sewer gases freely. These openings are usually so located and of such shape that it is a difficult matter to close them perfectly, and under such conditions the householder feels that he has absolutely no protection. In summer, when a house is to be closed, perhaps for weeks at a time, it is almost certain that evaporation will unseal traps, and then, in spite of plugs or other precautions, the house is open to the enemy. Some two years since I had occasion to examine the plumbing of a number of houses, the rent of which varied from $700 to $1000. I found that on the second floors neither basins nor bath tubs were provided with traps, and when the plugs were used to cut off the gases the then unexpected difficulty of the overflows presented itself. A short time since a friend of mine moved into a house plumbed in this style, and was soon compelled to invent something that would make the place safe to live in. The first thing done was to cover all the overflows. This was accomplished with putty in some cases, and in other cases with pieces of muslin put on with shellac varnish. The plugs were kept in all the time,

Overflows and safe wastes

Examples of bad workmanship.

Closing overflows.

WASTE AND SOIL PIPES. 55

except when removed to discharge waste water, and a little clean water drawn upon them to make certain that they were properly seated. By keeping the water-closet cover closed, in addition to frequent flushing, the house was rendered comparatively free from the faint, depressing odors of sewer gas. I have been thus particular in describing a case in point, as the means employed may be of service in similar instances. The plumber will at once think of the dangers attending the closing of all overflows, and it was not without some misgivings that the plan was adopted. The worst that can happen, however, is the destruction of walls and ceilings, and if due care is exercised this will never happen, especially if the whole household are aware of the new dangers attendant upon the closing of the overflows. These dangers are small as compared with those likely to arise from the introduction of sewer gas into the house.

In the best water-closets in the market, the weak place is generally found in the overflow. Were it not for their overflows, some of these would be absolutely impregnable to sewer gases under the heaviest pressures. Slight as the danger is from this source in good water-closets, and small as is the risk in cases where the plumbing is honestly and intelligently done, I think it is possible to arrange the plumbing work so that an escape of sewer gas into the house cannot take place through the overflows, at least. The plan is this: Make all overflows a system by themselves, allowing none to enter the soil pipe or its branches. Into the overflow system take also the safe wastes, which are all brought together and run into the cellar at some convenient point where they can be allowed to empty into a tank holding from, say, 10 to 20 gallons. In case any tub or basin runs over, the water at once finds its way to the tank in the basement or cellar. Here is, of course, perfect safety, so far as immunity from returning gases is concerned. There are some other precautions, however, to be taken. The outfall of the waste water must be so located that when running

Water-closet overflows.

How to provide for the waste of overflows and safes.

the noise shall be heard, or attract attention in some way. The water may in some cases be arranged so as to flow directly into the street, and thus escape the necessity of a sewer connection for the overflow pipes within the house. Where the tank is located within the cellar, it can be connected with the soil pipe by means of a pipe closed by a cock. This should be arranged by means of a float, so that the cock will be opened before the tank is full, and closed when it is empty or nearly so. Thus the sewer connection is entirely cut off, while at the same time the overflow water has a free escape. With a house thus arranged several things are possible which could not be obtained by the ordinary system. For example, if it is desired to leave the fixtures unused for any length of time, we can put in all the plugs in basins, sinks and tubs, and thus entirely exclude sewer gas, while no risk of overflowing is incurred. From water-closets provided with tight-seating valves, there is no danger save from the overflows which are sealed with water only, and where these overflows are arranged in the manner we have described, a house with such fixtures is safe for any length of time. Neither the breaking of seals by pressure nor their evaporation would cause any leakage of gas, the sewer being entirely cut off from the house.

Automatic flushing of safe and overflow wastes. In some very costly work I have seen arrangements for automatically flushing the safe wastes. They are supplied with water by a service-pipe branch, and in the case of water-closets the opening of the valve allows a quantity of water to pass into the safe and thence into the safe waste. This seems to me a clumsy method of accomplishing what should never be necessary. The overflow and safe waste are only provided to guard against accidents, and the water passing through them can be easily disposed of without giving these pipes a direct connection with the sewer.

A line of soil pipe extending up through a house should, if possible, be so placed as to be easily accessible on all sides. This is a matter for the architect to look after. A very common practice is to set them in niches, or recesses, built in the walls. To this some plumbers make no objection; but a majority of those who believe in doing good work, and know how to do it, condemn this disposition of the pipes in unmeasured terms. The plan is objectionable in some respects, though not so bad as to afford any excuse for slovenly workmanship on the part of the plumber. When soil pipes are placed in recesses, it is generally difficult to make good joints in them, though not impossible unless the space is very narrow and the pipe so large as to nearly or quite fill it. There is usually insufficient room for the use of the calking iron and mallet, and the plumber is therefore compelled to provide himself with special tools when he undertakes a job of this kind. When the pipe has to be put up in lengths, and joints made after the line is set in position, the careless plumber will usually content himself with running in hot lead and leaving it as it cools. One plumber of my acquaintance, who is both ingenious and honest, gets over the difficulty by leaving a lower joint open until the rest are done, which enables him to turn the whole line around and calk all sides equally. The plan is a good one, but architects cannot expect that plumbers will generally take this trouble. *Setting pipes in niches. Difficulty of calking. An expedient for calking in niches.*

Another and serious objection to this method of disposing of the waste pipes of houses, is found in the difficulty usually experienced in getting access to them when they need repair. The amount of space saved inside a building by such a disposition of the waste pipes is of no consequence to anybody. An architect who has any resources can always make provision in his plans for the pipe system, without disfiguring any apartment or sacrificing a square inch of space available for any other use. The pipes should be concealed by movable screens or panels of some sort, and should be accessible throughout their entire length. To build them into walls, and especially to cover them over with lath and plaster, so that they cannot possibly be inspected without breaking through the walls, is not *Importance of having pipes accessible. Movable screens or panels.*

good practice. The disposition made of them must, of course, depend upon the character of the building, its interior arrangement and the shape and ornamentation of the rooms. The problem is one which the architect must solve by the aid of a little common sense. As the soil and service pipes are usually carried up side by side, the position selected for them should be one in which there is least danger from frost in cold weather. Obviously, this will not be against the outside wall of the building. A sheltered corner in which they would be beyond the reach of frost, or an angle of the central partitions, can easily be found in any house which is not planned with an utter disregard of the fact that it is ever to be drained.

Disposition of pipe dependent on plan of house.

Protection from frost.

In good plumbing practice the soil pipe is always larger than its largest branch. In common work we often find them the same size. When the branches connecting water-closets with the soil pipe are 4 inches in diameter, the soil pipe should be at least 5 inches. For this there are two good reasons. Anything which is forced out of the closet traps should pass easily through the soil pipe, or we are liable to have obstructions at points beyond reach; and the soil pipe should be so large that it cannot run full under ordinary circumstances, even when we have water simultaneously discharged into it from several branches. The force of this second reason will be shown in the remarks on the syphoning of seals in the succeeding chapter.

Main wastes should be larger than branches.

An iron soil pipe should not be too light. In much of the cheap work of the time the pipe used is lighter than it should be. I have seen pipe set up in houses which, tested with callipers, I have found to be not more than one-eighth of an inch thick. The objections to this kind of pipe are numerous and important. It does not possess the requisite strength; it is too quickly eaten through by rust, and it is very apt to have sand holes and other imperfections which, for a time, may afford an easy outlet for the gases of the sewer. The difference in cost between light pipe and that of suitable thickness (a quarter of

Heavy pipe the best.

Objections to light pipe.

WASTE AND SOIL PIPES.

an inch for private houses, and three-eighths and upward where there is a long line of large size to accommodate a continuous outflow of considerable volume), is not great enough to make the economy profitable. In architects' specifications we seldom find a suitable weight of iron pipe called for. Consequently the principal demand is for very cheap and light pipes. As made they are as hard as chilled iron—owing to the fact that they are cast so thin—and about as brittle and difficult to cut as glass. If dropped they crack or break, and are utterly untrustworthy at all times. In much of the cheap work of the time we find 4-inch iron pipes used which average about eight pounds to the foot. In good work 4-inch iron pipes should be at least of 12 pounds weight to the foot. I know of one large public building in New York in which a 10-inch pipe weighing 45 pounds to the foot is used. *Weight of pipe for good work.*

Thus far I have considered only vertical lines of soil pipe, with their branches. In planning a drainage system for a house which was to be made as perfect as possible without regard to cost, I should have separate lines of soil pipe for each floor. I have known instances in which every waste pipe in the house was carried separately to the main waste in the cellar, and each was separately ventilated above the roof. This plan has the advantage of confining obstructions to the pipes in which they occur, without interfering with any of the others; but it involves extra expense, and cannot be considered indispensable to good drainage in average houses. *Soil pipes for each floor.*

In sewer-drained dwellings it is usually necessary to give the soil pipe nearly a quarter turn bend below all the house connections in order to reach the sewer, as kitchens, bath rooms and water-closets are commonly in the rear apartments, while the sewers are under the streets upon which the houses face. The pipe must, therefore, be carried along the cellar wall or under the cellar floor. The former is preferable when practicable; but in what are known as English basement houses, in which the kitchen is on a level with the cellar, the pipe which *Extension of soil pipe to sewer.*

receives the kitchen drainage must be laid in a trench under ground. Sometimes this section of the main waste pipe is made of a different material from the vertical line, usually because a cheaper one than iron can be had. This is an economy which does not pay in the end.

Under all circumstances I should recommend carrying the iron soil pipe all the way to the sewer, and in the best work of the time this plan is followed. I have never seen a house drain built of stone, brick or wood, and rarely one built of earthen pipes with cement joints, which I should be willing to live over. Stone drains, having rough inside surfaces, cannot be effectively flushed, and become coated throughout with foul deposits, offensive and dangerous in their rapid decomposition. Brick drains, as usually built, have this same objection, together with the liability of all but exceptionally good bricks to disintegrate when buried and kept constantly wet. Even when highly vitrified and laid with hydraulic cement, their rough surfaces and the perviousness of their joints to water are objections which should exclude them from use for this purpose. Earthen pipes, even when well glazed, cannot be depended upon when laid in cellars, for the reason that the best cement joints are pervious to water, which carries with it organic matter to lodge and decompose in the pores of the pipe and its joints. Wooden drains usually leak and always rot very rapidly. When made of plank, they cannot be given a shape favorable to effective cleansing. If set with the flat side down, the scouring power of the water flowing through them is reduced in proportion to the surface covered by it and the resistance it encounters from friction. If set cornerwise a better result is obtained, but the bottom of the V is liable, in the case of a house drain, to become filled with solid matter which resists the efforts of the water to carry it away. The next best thing to iron pipe is glazed Scotch drain pipe joined with Portland cement, but iron is so much better than any substitute yet found for it, that it should, I think, always be exclusively used in the drainage of city houses. The different methods employed in the drainage of country houses, generally render the use of earthen pipes desirable and economical.

Soil pipes carried through cellars should have a continuous support. If conducted along a cellar wall, which is the best practice under favorable conditions, they should be laid upon shelves of stout plank, which, in turn, are held up by suitable brackets. A few pipe hooks will keep them in place, but these should not be trusted to carry the weight of the pipe. If laid under a cellar floor, a strip of wood should be placed along the bottom of the trench, which will give them a uniform descent to the street. When iron pipe is used this is very desirable; with earthen pipe it is absolutely indispensable. In digging a trench it is scarcely possible to preserve the grade so perfectly that a pipe can be laid in it without sags. These sags are so slight, especially when the grade is easy, that they readily escape notice, but they are almost always great enough to retard the flow of water and diminish the scouring power. The air in the pipe will usually find a lodgment at the highest points along the line, and when the pipe is trapped near the entrance to the sewer (as I think it never should be), these accumulations of air are likely to offer a very effective resistance to the passage of water. With a very steep grade slight sags do not usually give trouble, but it is desirable to avoid them in house drains. A tendency to sag may break or loosen a joint and cause a leak which, if under ground, may escape notice until many cubic yards of soil are saturated with foul water, and a nuisance is created which can only be abated by excavating the place and filling in with fresh earth.

Sections of soil pipe carried under houses should have all the fall which can be conveniently given them. Nothing is gained and nothing saved by flattening the grade. If a pipe is carried along a side wall, an ample fall can always be secured; when laid under the cellar floor, it should have all the fall possible, which is not always as much as would be desirable. The aim of the architect should be to give that portion of a soil pipe which deflects from the perpendicular as steep a descent as possible. A pipe cannot be too completely scoured by the water passing through it.

The opening in the foundation wall of a house for the soil pipe to pass through should always be enough larger than the outside diameter of the pipe to allow the walls to settle without crushing or deflecting it. When this precaution is neglected, the pipe is liable to be broken or disjointed by the settling of the foundations, allowing the sewage to escape and work its way back into the cellar. I know an instance in which a pipe was thus broken, and the owner of the house, ignorant or careless of the fact, neglected the necessary repairs until sickness made its appearance in the family, when his physician, a very practical man who liked to know the reasons for things, made an inspection on his own account and discovered the source of the mischief.

So far as this part of the subject is concerned, it only remains for me to urge upon those connected with the building trades, and upon physicians, house owners and tenants, the vital importance of good workmanship in all departments of plumbing work related to or connected with house drainage. In this chapter and the chapter following on Traps and Seals (to which I refer the casual reader), I have endeavored to embody such plain directions concerning waste and soil pipes as I have considered necessary to give a person of average intelligence an idea of the conditions under which it is possible to secure immunity from dangers that, in cheaply piped houses, are almost certain to attend the opening of connections with the sewers. Without going into details of interest only to plumbers, who ought to be familiar with them, I have endeavored to describe the means by which a line of soil pipe can be put up so as to be securely held in place and have tight joints and connections throughout. There are, doubtless, other ways of accomplishing the ends sought, but I know of none better or more certain

WASTE AND SOIL PIPES.

under all circumstances than those of which I have spoken. During the past few years I have inspected much of the best plumbing work of the time, and a great deal of work which was not good at all, and the conclusion I have reached is that when sewer gas finds its way into a house through the soil and waste pipes, the fault lies somewhere between the architect, the builder and the plumber. In any case it is without excuse. I know that houses can be drained into sewers—and even into foul and unventilated sewers—without bringing sewer gas into them. The existence of foul sewers is in itself a perpetual danger to the public health, but there is no reason why we should bring that danger into our houses by providing channels through which the poisonous air of the sewer can find a means of ingress. I know of houses into which no sewer gas ever comes—unless, possibly, through the windows, borne in with air of the street—and I have no hesitation in saying that, when the tenants of houses demand immunity from the dangers of unhealthful conditions, architects and builders will find a means of correcting the evils now complained of as practically irremediable. Sanitary reform in cities only waits until those to be benefited by it shall demand it.

Responsibility for sewer-gas poisoning.

Immunity possible.

CHAPTER IV.

TRAPS AND SEALS, AND THE VENTILATION OF WASTE PIPES.

Traps in waste pipes. The importance of trapping all waste pipes, and of maintaining in every trap a seal of water, is probably better understood and appreciated than any other essential of safety in the drainage of houses. There is, however, a great deal to be said concerning traps and seals which, if known and applied in plumbing practice, would have a most important influence upon the public health.

Too much expected of traps. The service expected of water seals in traps of every form is usually much greater than they are capable of performing. For this reason undue reliance is placed upon them. Those who suppose that a small quantity of water held in a bend of a waste pipe is alone sufficient to prevent an inflow of sewer gas through that pipe, know very little about either water or sewer gas. We can only accomplish this most desirable result *The security of traps dependent on ventilation.* when we afford the sewer gas an easier means of escape than it finds through the water in traps, by ventilating the soil pipe and its branches. In common work the waste-pipe system usually ends in the highest water-closet or wash basin in the house. To prevent the escape of gases generated in the sewer, and which naturally find their way into the pipes under certain conditions to be noticed further on, dependence is placed upon the water *Popular errors respecting water seals.* seals in the waste pipe traps—the common argument being that, as the express purpose of the seal is to prevent the escape of poisonous gases and foul odors, other means to this end will be superfluous. This argument is based upon an evident mis_ understanding of the conditions which exist in sewers. In New York, for example, there are periods, often of several weeks' *Unventilated sewers.* duration, when the sewers are absolutely without ventilation, and when the only escape for gases generated therein, often held

TRAPS, SEALS AND VENTS. 65

under considerable pressure, is through soil and waste pipes. Here, as in most American cities, the chief dependence for sewer ventilation is upon perforations in the manhole covers. These are better than no openings at all, when they are open, but they are very liable—certain, indeed—to become choked with mud and dust during much of the time; and from the first heavy snowfall of winter until spring—with, perhaps, a few brief intervals of general thaw—they are as effectually closed by ice and snow as they would be if covered over with the permanent pavement of the street. The culverts at the street corners are of course trapped, and during the winter season they are likely to remain effectually sealed. The mouths of the sewers are, as the rule, so placed as to be completely submerged at high tide, at which times the river water forces its way up into them for a considerable distance, compressing the air confined within in proportion to the resistance offered at the various outlets by which it seeks to escape. To increase this pressure we have still another active agent—heat. In cold or cool weather the temperature of the air in sewers is considerably above that of the outer air. We are continually pouring great floods of hot water into them, at temperatures ranging from 80° to 180° Fahr. It is not unusual to allow steam engines to exhaust into them, and, as showing that the temperature of the confined air of sewers is not low enough in average weather to condense steam, I may instance what may be seen any day in the streets of New York—the escape of steam, still a hot vapor, from the perforations in manhole covers in regular puffs corresponding to the piston strokes of an engine in some neighboring building. In one way or another we impart considerable heat to our sewage besides that generated in the process of decomposition. In very warm weather the temperature of the outside air is usually higher than that of the air in sewers; and, so far as unsealing traps is concerned, a difference in the temperature of the air inside and outside a sewer is nearly as effective one way as the other. A simple experiment,

Manholes.

Mud and snow in the streets.

Tide water in sewers.

Expansion of air in sewers by heat.

Temperature of sewers.

Effect of changes of temperature in sewers upon traps.

5

described by Latham, will serve to show how such differences of temperature will affect the security of a water seal in a trap.

Experimental demonstration. In the illustration marked Fig. 1, is shown a glass flask with a bent glass tube inserted in the cork, the bend forming a trap which is filled with water. If the hand is placed on the flask, its warmth is sufficient to so expand the air within that enough of the water in the bend of the tube is driven out to leave the trap unsealed. By partly immersing the flask in cold water, the air within it is so contracted in volume that the pressure of the outside air forces the water in the tube into the flask, also effectually and promptly unsealing it. The air in the waste pipes of an occupied house is daily subjected to frequent expansions and contractions, which may, *Unsealing traps.* and often do, unseal traps. Under these conditions it is readily seen that when the mouths and manhole ventilators of our sewers are closed, any increase in the volume or temperature of the flow will cause the confined air to struggle for a means of *Pressure.* escape, which it usually finds at some trap. It will do this just so soon as the pressure equals that of the column of water in the trap.

Fig. 1.

Resistance of seals to pressure. To displace a seal altogether, no very great force is necessary. A seal 3 inches deep offers a resistance to the passage of air equal only to a pressure of two ounces per square inch. With adequate ventilation of the soil pipe, the pressure upon waste-pipe traps is relieved, and there is little danger of sewer gas find-*Benefits of ventilation for waste pipes.* ing its way into a house when such ventilation is provided— presuming, of course, that the plumbing work is judiciously planned and properly executed in other respects.

To the pressure within the sewers, tending to displace the *Draught in traps.* water seals of unventilated waste-pipe systems and to force sewer gas into our dwellings, we must add what is commonly called the "suction" of the house. During much of the year, and especially during the season when we live with closed windows and doors and depend on fires, there is a constant outflow from the house through the chimneys, producing a partial vacuum which varies greatly under different circumstances. In any case it is enough to cause a great deal of air to enter by a very small opening, and when doors and windows are carefully sealed with list and weather strips, the draft upon the traps may be very considerable. Were the traps empty there would be a strong and steady inflow of air through the pipes; and when they are sealed the "suction," as it is called, very effectually supplements the pressure resulting from the causes already noted. *Rarification of the air in houses.*

In trapping pipes there is a good chance for the unskillful plumber to defeat the end he seeks to accomplish. My attention was lately called to the house of a friend in which the faint, depressing odor always noticeable in the bath room and adjoining apartments revealed the presence of sewer gas. A thorough examination of the plumbing work showed certain defects only too frequently met with in badly piped houses. In the bath room there was a tub, water-closet and wash basin, after the usual practice, but the bath and basin wastes were conducted into the soil-pipe branch below the water-closet trap. These pipes were trapped, but the bath overflow was not, and through this the sewer gas found easy access. Stopping the overflow remedied the difficulty in part, but not wholly, as the same smell was noticed until the waste-pipe system of the house was thoroughly ventilated. Had all the pipes been trapped and every trap been full of water, the house would not have been a safe one to live in until their ventilation had been provided for. In another instance, when inspecting an extensive pipe system in a large hotel in a Western city, I found that an attempt had been made to form a stench trap in every waste pipe, but in a great many instances the pipe had been bent at such an angle that the water held in the bends would barely make a seal, while others were so long and so nearly horizontal that the force of the water coming down the pipe would, in all *Defective trapping. Examples of bad workmanship.*

probability, carry it past the bend and leave the traps unsealed. Evidently the traps were home made, as no two were alike and not half of them were well made. It may be assumed that no trap can be so sealed with water as to offer an effective barrier to the passage of gases held in an unventilated waste pipe, whether under pressure or not.

As further illustrating the importance of ventilating the waste-pipe system of a house, I may mention a phenomenon which has surprised a great many people, practical plumbers included. It sometimes happens that traps which are to all appearance properly set, so nearly empty themselves as to afford a free passage for air through them. This may occur in several ways. For example, if we have a quantity of water thrown into the trap so great as to fill the body of the pipe for a distance greater than the vertical hight of the column of water in the trap, it will be quickly unsealed. This action is easily explained, since it is only a syphon in a slightly modified form. When we take a tube bent into the shape of the letter U, fill it with water and invert it, there is no tendency on the part of the water contained in it to run from one end of the tube or syphon more than from the other, so long as the two legs are of equal length. The liquid in both legs may be said to be pulling downward and tending to form a vacuum at the highest point of the bend; being of equal weight they balance each other. If we lengthen one of these legs, the water in the long leg will be heavier than that in the short leg, and will run out, while atmospheric pressure will force the water in the short leg up to the top and out of the tube. Now, if the short end of the syphon be dipped in a vessel containing water, as shown in Fig. 2, the well-known

Emptying of traps by syphoning.

Causes.

Phenomena of the syphon.

Fig. 2.

TRAPS, SEALS AND VENTS.

action will take place of emptying the vessel by syphoning. Applying this illustration to the arrangement of waste pipes and traps shown in Fig. 3, we find that if the flow of water, solid matter, paper or anything of the sort from the water-closet b is great enough to fill the pipe a, we have a column of water flowing down which is heavier than that in c, and tending to form a vacuum above it in the main pipe. Consequently, the water in c and d will be forced over into the pipe a by the atmospheric pressure. Not only will this happen in the trap f, but at the same time, if there is a heavy flow of water down the soil pipe a, as from the bath tub k, and no ventilation, the traps g and h may be partly emptied, leaving a free passage for foul air from the sewer until they are again filled. The emptying of traps may also occur from the creation of a vacuum in a soil-pipe branch, as at c, by water, as from d, flowing past the orifice at which it discharges.

How syphoning may occur in waste pipe traps.

In the drawing marked Fig. 4, I have shown an arrangement of the soil pipe and its principal branches which I consider free from any such objections. It of course admits of such extension as may be necessary to adapt it to the various styles of architecture employed in cities. In this arrangement the soil pipe a is carried

Arrangement of waste pipes.

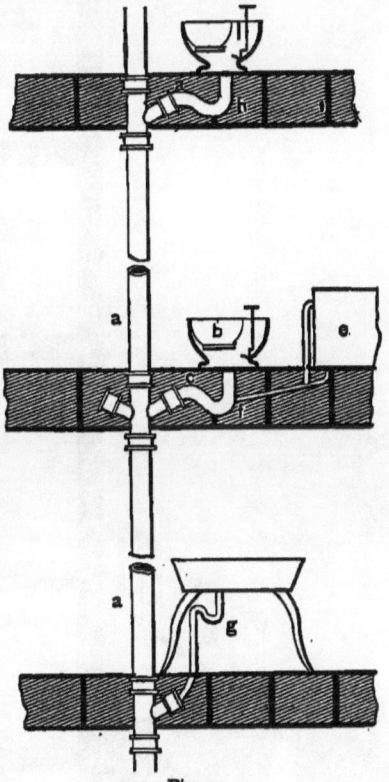

Fig. 3.

70 TRAPS, SEALS AND VENTS.

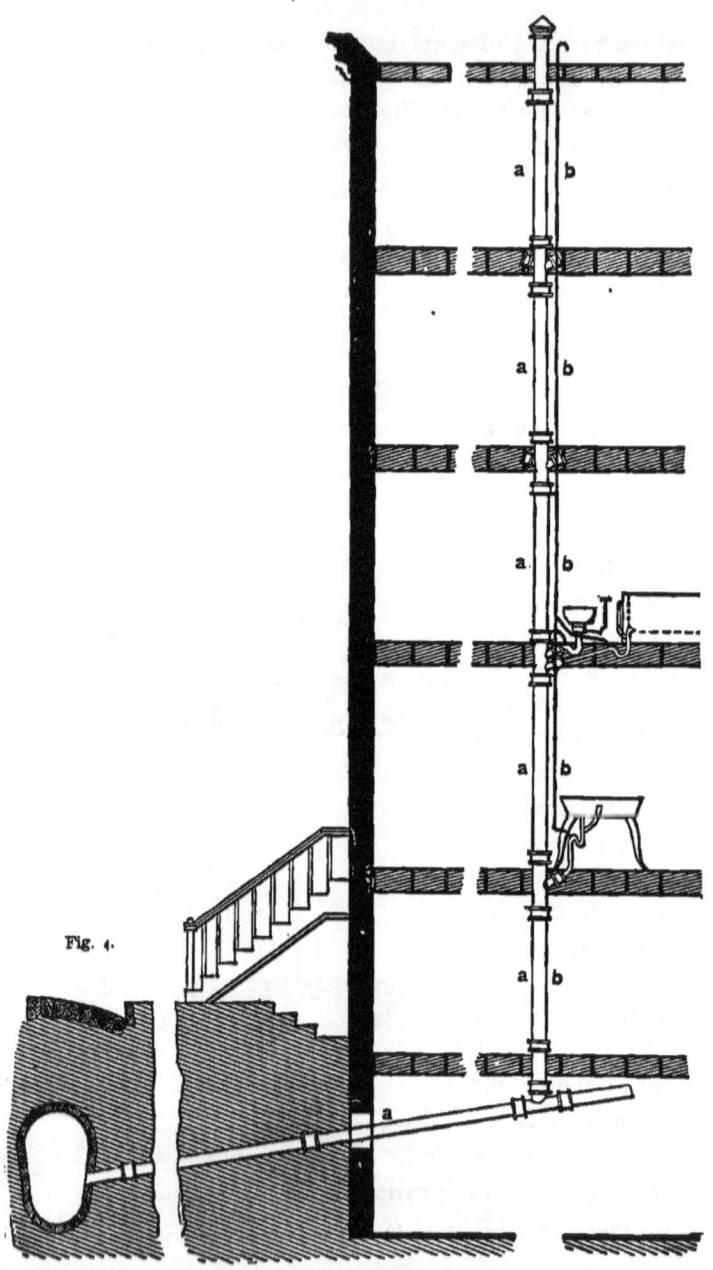

Fig. 4.

TRAPS, SEALS AND VENTS. 71

straight from the sewer to the point at which it takes the upward bend. This section is of the same material as the vertical line, and is without a trap at any point. From the bend in the cellar it is carried vertically up to and through the roof, terminating in a ventilating cap. This is no doubt unnecessary, but when the pipe is not bent over, some form of cap is commonly used for the purpose of keeping the wind and rain out. To leave the pipe open at the top answers every purpose. Parallel with the soil pipe, also following it to the roof, is a ventilating pipe, b, which connects with every trap in the branch wastes and ventilates each one independently of the ventilation secured through the soil pipe. This pipe b, for the independent ventilation of traps, is seldom introduced, except in the very best class of work, but as it entails little increased expense, I should recommend its employment in every case. *No trap in the main waste pipe. The ventilating oxtension above the roof. Supplementary ventilation for traps*

It will not do to place too much dependence upon the ventilation of the soil pipe alone to prevent the emptying of waste pipe traps. It often happens that the waste pipes of wash basins placed at a distance from the soil pipe, syphon their traps even when the latter is ventilated. In such cases the wastes of basins, sinks or closets remote from the soil pipe, should be carried up to and through the roof. This is the easiest and surest way to secure good ventilation, and when such ventilation is neglected we usually have trouble from "the smell of the sewers," even though the main wastes into which these branches discharge are ventilated. When this precaution is neglected, the traps in these long branch wastes are very liable to be unsealed by syphoning. The pipe usually has a capacity for discharging more water than can flow into it, and the rapid downward current makes a partial vacuum which often causes the air to draw down through a considerable body of water to fill it. Every one has noticed, in emptying a bath tub or wash basin, that air is sometimes sucked into the waste pipe with a roar long before the tub or basin is empty. The same thing takes place after the water has all run into the pipe. *Syphoning of traps in long branch wastes. Ventilation for branch waste pipes. Vacuum in pipes caused by rapid flow of water.*

The rapid outflow creates a partial vacuum, which is more easily filled by an air current moving with the water than by one flowing against it, and the air rushing down the pipe will often displace the seal in the trap, leaving it open, though perhaps not quite empty. Ventilating the soil pipe will do much to protect waste-pipe seals, but no trap is absolutely secure against being unsealed by the several causes mentioned, unless separately ventilated as shown in Fig. 4. When this is done, the syphoning of seals in traps is an impossibility.

<small>No unventilated trap safe.</small>

The absence in Fig. 4 of any trap in the main waste or soil pipe will, no doubt, surprise a good many readers. In common practice this pipe is often trapped just inside the cellar wall, the idea being that sewer gas should be prevented from entering the house drain at all, and that a running trap will accomplish this object. In my judgment such a trap does vastly more harm than good. It is very liable to become choked by accumulations of solid matter, and unless well protected its seal may freeze in very cold weather, which is a troublesome matter. The most serious objection to it, however, lies in the fact that it detracts from the efficiency of any arrangements which may be made for ventilating the soil pipe. When we trap this pipe at any point we immediately close it against the passage of air currents, and the pipe above and below the trap is filled with what is known among plumbers as "dead air." When there is a copious dash of water down the pipe, as when a pail of suds is thrown into a slop hopper or water-closet on one of the upper floors, the stagnant foul air in the pipe must be displaced. The resistance offered at the trap in the soil pipe will generally be greater, owing to the larger body of water held in it, than at one of the traps in the lower branches, and the displaced air is very likely to blow through one or more of these.

<small>Traps in main wastes.</small>
<small>Objections.</small>
<small>Blowing through seals in branch wastes.</small>

It should be remembered that the most important of the ends sought by ventilation is the establishment of a free communication between the sewer and the outer air. When the pressure upon the air confined in the sewer is increased from any cause, it should have an outflow through every house drain. When from any cause a partial vacuum is created in the sewer, every house drain should be an inlet for air. In other words, we should allow the sewers to "breathe" through the main waste pipe of every house, besides giving them as many breathing holes in addition as can be provided. When every branch waste pipe is trapped, and especially if every trap is ventilated as shown in Fig. 4, this inflow and outflow of air through the soil pipe of a house is attended with absolute immunity from inconvenience or danger to health, provided the jointing is properly done. There is no pressure of foul air upon the traps, but little chance for water seals to absorb foul gases, and none at all for the unsealing of traps by blowing out or syphoning out. With such an arrangement we have nothing to fear as regards the security of water seals except evaporation, and this is easily guarded against. Should a seal be allowed to dry out, however, the danger of atmospheric pollution is much reduced by the double ventilation described. *House drains should ventilate sewers. No sanitary objections to the passage of sewer gas through ventilated soil pipes. Evaporation of water in traps.*

It is probable that many practical plumbers will consider the independent ventilation of traps unnecessary. I have seen many very satisfactory jobs of plumbing in which it was omitted, but for the reasons already set forth it is certainly desirable if not essential. Sewer gas is a subtle enemy to health, scattering broadcast the germs of fever and the seeds of infection, and when we undertake to securely intrench ourselves against its attacks we cannot afford to neglect any precautions of safety to save the cost of a few pounds of lead pipe or a few dollars for additional labor. *Independent ventilation for traps. Security better than economy.*

Where there is no danger of an escape of sewer gas to be apprehended, the ventilation of the waste-pipe system of a house would be necessary to prevent the rapid corrosion of lead wastes. The points at which lead wastes soonest wear out and begin to leak are the points at which, in the absence of ventilation, the lighter gases from the sewer would naturally accumu- *Ventilation prevents rapid corrosion in lead pipes.*

late. I find this is the usual experience of plumbers in this country, and the same phenomenon seems to have attracted attention abroad. At a meeting of the Society of Medical Officers of Health in London, held a short time previous to this writing, Dr. Andrew Fergus, President of the Faculty of Physicians and Surgeons of Glasgow, stated that while the water supply of that city, drawn from Loch Katrine, was so pure in itself and so distributed as to preclude the possibility of excremental pollution, typhoid fever had increased to a startling extent, and as the cause of this disease must be sought either in polluted water or in air poisoned by the gaseous products of decomposition, it was a safe conclusion that the latter was the cause of the trouble in Glasgow. Reciting many remarkable facts in connection with cases of typhoid fever and diphtheria which had come under his notice, he showed several pieces of 4-inch lead pipe curiously perforated by corrosion commencing from within, and which could only have been caused by the action of sewer gas. These perforations had admitted sewer gas into the houses in which mysterious cases of typhoid and diphtheria had occurred; and as similar perforations were not found in such portions of the pipes as were at any time charged with liquid, they were not betrayed by the leakage of water. Dr. Fergus further said he had found that such pipes as were freely open to the sky by an upcast shaft, remained sound nearly twice as long as those in which there was no effective ventilation, but he did not think that a lead soil pipe exposed to the action of sewer gas could be depended upon as sound for longer than 10 or 12 years. He also asserted, as the result of careful and intelligent observation, that the usual method of depending upon water seals in traps allowed sewer gas to diffuse itself through a house "by a process of soakage," and that in from half an hour to two hours the foul gases of the sewer and house drain would have saturated a seal, and thenceforth be freely emitted into the house. Other eminent and experienced physicians stated that they had reached the same conclusions

Typhoid fever in Glasgow.

Examples of internal corrosion of pipes by sewer gas.

Passage of sewer gas through water seals.

from observations made in London and other British cities, showing that the rapid corrosion and perforation of unventilated waste pipes is a matter of common experience on both sides of the ocean One of the many examples of the corrosion of lead pipe which I have seen came to my notice quite recently, and is especially interesting as showing how the defective drainage of one house may affect the healthfulness of another. I was consulted as to the cause of a very strong smell of sewer gas in the well-appointed residence of a gentleman of wealth, on one of the finest avenues of New York. This smell was chiefly noticeable in a passage connecting the front and back bedrooms on one of the upper floors. A careful inspection of the plumbing work of the house was made without discovering any serious defect. The bath room, water-closets, wash basins, &c., were all on the north side of the house, while the offensive smell was confined to a passageway on the south side, and this was not supplied with water. The first conclusion was that the sewer gas escaped from some defect in one of the pipes on the other side of the house, and passing under the floor found a means of escape through a crack in the passageway floor. A further examination showed that this was not the case, and it then became evident that the cause of the trouble was outside of the house. An examination of the adjoining building was then made, and the source of the trouble was easily found. The lead soil pipe, which was carried up against the party-wall, extended a few inches above the branch connecting it with the third story water-closet, and was closed by means of a lead cap soldered on. From the cap down, for 12 or 15 inches, the pipe was completely honeycombed. There were countless holes in it, from the size of a pin point to an eighth of an inch in diameter, and the shell was so thin and rotten that it could be readily crushed between the fingers. From the holes in the pipe, which was wholly unventilated, sewer gas had freely escaped, and had found its way into the adjoining house through the brick wall. The damaged pipe—which had clearly been cor-

Curious instance of atmospheric contamination by sewer gas.

Mistaken conclusions.

A corroded lead soil pipe.

Transmission of sewer gas through a brick wall.

roded by sewer gas alone, as no water had ever come into it—was repaired and the nuisance ceased. Those in the house in which the corroded pipe was found claimed to have suffered no inconvenience from the smell, which one would naturally suppose would have been intolerable to a person with an ordinarily sensitive nose, while in the adjoining house it was a constant annoyance.

Dangers which menace water seals in traps. As already stated, water seals in traps are liable to two dangers which are commonly overlooked, but either of which, in the absence of abundant ventilation, will wholly destroy their efficiency in a very short time. *Evaporation.* They may be reduced in volume by evaporation until a passage is left for air through the traps; *Saturation by foul gases.* or exposed for even a short time to contact with foul gases having no other means of escape, they will absorb and transmit them. *Frequent flushing.* The first of these dangers can, as I have said, be guarded against by frequently flushing the pipes. A careful housekeeper should see that this is done daily, at least; and when houses are closed or left in the care of servants during the absence of the occupants, provision should always be made for keeping the traps filled with clean water. This is commonly neglected, but the consequences of such neglect may be the ruin of health or the loss of precious lives. The danger of saturation by, and transmission of, sewer gas through water seals, is less readily apparent and calls for brief explanation. It will be found in the chapter on water-closets. Here we will deal only with the *Absorption of gases by water in traps.* fact, which is that water standing in the traps of unventilated waste pipes is constantly absorbing more gas than it can hold. Long before the point of saturation is reached, it begins to give off on one side the gases it takes in on the other. When the point of saturation is reached depends somewhat upon circumstances which need not here be taken into account; but it is a fact within the observation of all who have studied the subject experimentally, that water exposed to contact with confined sewer gas grows fouler and fouler, and never seems to wholly lose its capacity for absorbing until it becomes putrid and decomposition takes place within it. In an unventilated trap a water seal cannot remain pure for many hours. It begins to absorb at once, and the length of time required to so charge it that it will begin to give off gases is, probably, as stated by Dr. Fergus, from half an hour to two hours, according to the pressure of the gases in the pipe and their foulness. For this reason, every waste pipe, even when ventilated, should have enough water passed through it daily to completely replace the seal in its trap. When the pipes are unventilated, it should be done very often. *Renewal of seals.*

Dr. Fergus, whose eminence as a sanitarian justifies the liberal reference I have made to his writings, gives, in a contribution to the *Sanitary Record*, the results of some interesting experiments with traps and seals. He shows that the water seal of an ordinary S trap is scarcely to be regarded at all as a barrier to the passage of sewer gas, and that it has the further disadvantage of retaining enough decomposing matter to produce within the pipe itself a sufficient amount of sewer gas to accomplish all the evil that is feared, except such as comes by the conveyance from the sewer itself of gases containing the germs of disease originating in other houses. He made a series of experiments with a bent tube, its bend being filled with water after the manner of the usual trap. In the sewer end of the tube he inserted a small vessel containing a solution of ammonia. In fifteen minutes the ammonia had passed through the water of the trap, and had bleached the colored litmus paper exposed at the house end. In another experiment he produced the rapid corrosion of a metal wire exposed at the house end. To prove that this transfusion takes place not with ammonia alone, which is lighter than air, he made the same experiment with sulphurous acid, sulphuretted hydrogen, chlorine and carbonic acid, all of which were transmitted so as to produce their chemical effect on the other side of the trap within from one to four hours. He says: "We are, therefore, very strongly inclined to believe the last alternative, namely, that however *Dr. Fergus's experiment with traps and water seals.* *Gases passed through water.* *No safety in traps.*

well drains may be trapped, sewer gas will find its way from them into our houses, and any one who is acquainted with Graham's investigations as to the diffusion of gases, will readily understand how this will happen."

Ventilated traps. The same experiments were made with a ventilated pipe, and although the action was slower, it still took place. The ventilation was not such as to insure a clear and direct flow of air through the pipe; but it was more effective than many devices in use, and was considered quite sufficient.

Concerning the decomposition taking place within the trap, Dr. Fergus believes that it is almost impossible, even with a copious flow of water, to cause undecomposed fæces to be discharged through the bend. With any ordinary flow there is only an eddying of the water in the trap, not a sufficient movement of the whole volume to carry floating matters under the bend. **Lodgment of fæces in traps.** "No more flow of water will carry out the fæces, which simply kept whirling round in it." Often, with the mistaken idea that by adding a few inches to the depth of water in the trap an effective resistance will be opposed to the pressure of sewer gas, these are made quite deep. **Shallow traps preferable.** Dr. Fergus recommends that the dip or bend should be only sufficient to secure a sealing, for the deeper it is made the more complete will be the retention of decomposing matters at the house side of the trap. **Disinfection impracticable.** In view of the large amount of water being discharged, he considers it practically impossible to use any disinfectant in sufficient proportion and with sufficient regularity to secure a chemical disinfection of the suspended organic matters.

My own observations and experiences lead me to conclusions which generally accord with those expressed by Dr. Fergus; but if I may be permitted to criticise the statements of so eminent an authority, I will say that he seems to have overlooked the fact that, with adequate ventilation, the protection afforded **Conditions of security for water seals.** by a water seal, renewed as often as once a day, is enough to prevent sewer gas from entering a house through a trapped branch waste. I know that such is the fact in this country, and

TRAPS, SEALS AND VENTS.   79

if not in Great Britain it must be because different methods are there followed in plumbing houses. It will be noticed that Dr. Fergus admits that the venting of the pipes experimented on was not such as to insure a clear and direct flow of air through the pipe. This is the one essential condition of good ventilation, and with this insured a trap with a half-inch seal which cannot be syphoned out may be depended upon. I would, however, urge the advantage of occasionally—if possible, daily—flushing waste pipes from basins, &c., not in constant use. Good ventilation of primo importance.

In a work of this kind, which deals with principles rather than with the details of plumbing practice, it is scarcely necessary to describe the several forms of traps in use. They are of two general kinds—those sealed by water and those sealed by valves. The latter, though somewhat extensively employed in foreign plumbing practice, are but little used in this country, and never in good work. Some modern varieties are still kept for sale by dealers in plumbers' supplies, but they are not to be depended upon for use in connection with house drainage. In all water-seal traps the principle is the same, whatever their shape, namely, a bend in or chamber attached to the pipe filled with water, which cuts off the passage of air but offers to liquids and semi-solids passing down the pipe no greater resistance than that due to the bend. As regards resistance to the *pressure* of air or gas, traps are efficient in proportion to the *depth* of seal they contain. Consequently, the volume of water in the trap is of no Forms of traps.
Valve traps.
Water traps.
Resistance to pressure in proportion to depth.

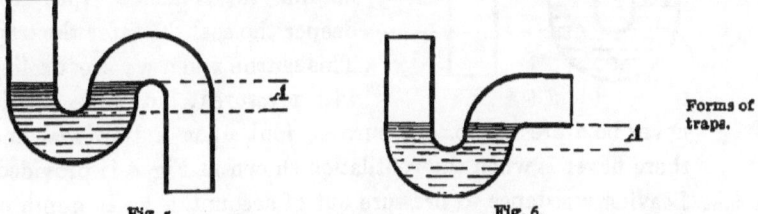

Fig. 5.   Fig. 6.

Forms of traps.

importance as affecting the depth of the seal. In the accompanying illustrations are shown three forms of traps in common use. The first of these, most commonly employed, is the S Depth cf seal not dependent upon capacity of trap.

trap (Fig. 5). The second (Fig. 6) is a form of trap in general use, embodying the same principle. Fig. 7 is a trap of the flat syphon variety, sometimes used, but of very limited utility in connection with house drainage. In each of these the distance, A, between the dotted horizontal lines projected to the right, shows the depth of the seal. Were the form of trap shown in Fig. 5 changed to the form shown in Fig. 8, it is obvious that the volume of water it holds would be very much increased, but the depth of seal and the efficiency of the trap as regards resistance to pressure of air or gases, would remain unchanged. On the other hand, were the trap shown in Fig. 5 changed to that shown in Fig. 9, we should have a less volume of water than in Fig. 8, but an increased depth of seal. A pressure of air which would blow through the seal shown in Fig. 5, would also blow through that shown in Fig. 8, but a considerably increased pressure would be required to blow through that shown in Fig. 9. Arguing from this fact alone, it would be natural to conclude that the deeper the seal the safer the trap. This is true when we are dealing with pressures; but there should never be a pressure of air, pure or foul, upon a trap seal, and there never is when the ventilation shown in Fig. 4 is provided. Leaving resistance to pressure out of account, a great depth of seal is unnecessary. With adequate ventilation, the service which seals are expected to render is limited to closing waste

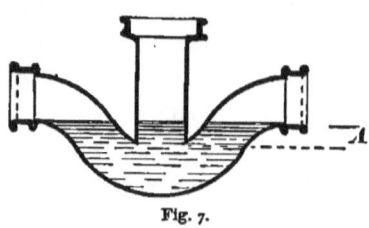

Fig. 7.

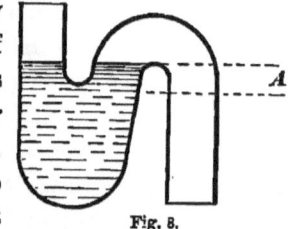

Fig. 8.

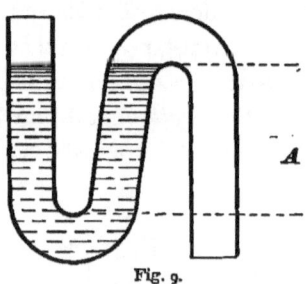

Fig. 9.

Dip as distinguished from capacity.

Much dip unnecessary.

TRAPS, SEALS AND VENTS. 81

pipes opening into houses against a free inflow of air through them. Without ventilation we should not be safe with traps having a depth of 12 or 15 inches, for while their seals might resist a considerable pressure (in the case of a 12-inch seal about 8 ounces per square inch of sectional area), they would absorb and transmit sewer gas as readily, though perhaps not as rapidly, as those held in pipes of much less dip. We must also consider the resistance to flow offered by a trap of any form, and the danger of obstructions resulting from the accumulation of solid and insoluble matter therein. When we give a pipe even 3 inches of dip, we are liable to have trouble from this cause without gaining increased security. From half an inch to an inch dip is the most we find provided for in the best modern plumbing practice. Many practical plumbers of experience consider half an inch dip as good as an inch. This is probably true, but as a safeguard against rapid unsealing by evaporation, I should recommend a dip of not less than an inch under average conditions.

Pressure required to displace seals.

Accumulation of matter in traps.

An inch dip required under average conditions.

Two of the best forms of traps I have ever seen are shown in

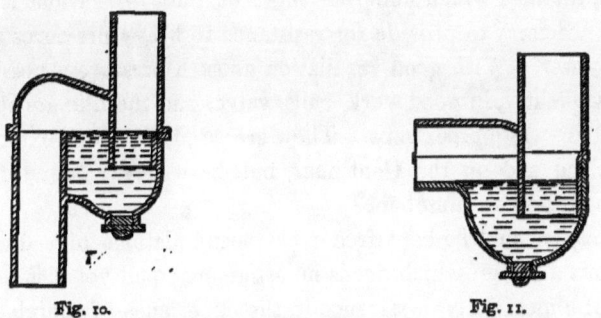

Fig. 10.   Fig. 11.

Figs. 10 and 11. One great advantage of these traps over those of the ordinary form lies in the fact that they cannot be syphoned out so as to break the seal. A heavy rush of water, producing a vacuum in the trap, displaces a part of the seal, but the amount of space in the body of the trap is so great and of such a shape, that the air passes through the water without driving the water

Traps which cannot be syphoned.

6

in front of it in a body. In other words, there is a large volume of water to be lifted up through the trap, but the stream of air which can be made available to force this water out is quite small, and hence passes up through the water while the heavy body of water falls back on each side, instead of being carried over into the waste pipe. I have repeatedly attempted to syphon out these traps, but there was always water enough remaining, after the heaviest flow, to form a seal. The importance of this feature is one that can hardly be overestimated.

Experimental tests. My experiments were made with traps in the sides of which watch crystals had been inserted to give a full view of what was going on within. The operation of the trap is thus easily seen, and the difference of action between it and the ordinary S or half-S trap plainly shown. The very compact form of *Convenience.* this trap enables the plumber to get a large one into very small space, an advantage which is important when it is desired to make a neat job.

The number and variety of new traps patented every year would require a separate volume to describe them. As the rule, the idea which underlies most of these inventions is that it is necessary to provide for resistance to heavy pressures from the sewer. With good ventilation no such pressure is possible; consequently, in good work, balls, valves and the like are rarely used in waste-pipe traps. They are used to some extent in England and on the Continent, but have never found favor with American plumbers.

Sediment in traps Traps should be kept free from accumulations of sediment. This is a truism which needs no argument; and yet it is a matter of almost daily occurrence in the experience of plumbers to be called upon to clean out traps which have been closed by articles carelessly or viciously thrown into water-closets and which could never pass into the sewer. When he suggests that the cause of an obstruction will be found to be some bulky article lodged in the trap and which has no business there, he is invariably answered: "Oh, no; that is impossible. We are always very careful not to allow anything of the sort to be thrown in." Upon investigation he may be pretty sure of finding a miscellaneous assortment of lost articles. In the house of one of our most experienced New York plumbers, a scrubbing brush and several smaller articles were taken from a water-closet trap, and not long since I was told by a Brooklyn plumber that he had found in a water-closet trap a valuable watch and chain and other articles, supposed to have been stolen by one of the servants. Similar instances might be given without number. Children and servants seem to be chiefly responsible for the obstruction of traps. It is no unusual thing for plumbers to find lost articles of jewelry, laces, house cloths, old shoes and worn-out rubbers, combs and brushes, broken crockery and mantel ornaments, broken glass and numberless other articles, which dishonest or careless servants or mischievous children have thrown in, ignorant of the fact that traps are great conservators of rubbish, and almost always retain whatever of solid or insoluble matter goes into them. I have known instances in which much trouble has resulted from the fact that servants would persist in throwing coal ashes into water-closets to save themselves the trouble of carrying them down stairs. It is unnecessary to say that the throwing into a trap of anything not intended to go into it, is wrong. In water-closets care should even be taken that stiff paper which resists reduction to pulp, or large pieces of newspaper, shall never enter them. Tough paper, or large wads of it, are likely to find a lodgment somewhere in the pipe, and, if they do nothing worse, will diminish the scouring power of water passing down it. Rags are especially troublesome in this respect, and when caught upon a projecting roughness, will sometimes cling for a long time.

The trap screw is a device for emptying and cleaning traps which has lately come into very general use—probably because architects have got into the way of calling for traps with trap screws in their specifications. Sometimes, for convenience, the plumber puts the trap in so that the screw comes on top—at least I have seen such cases—but as a general thing this makes no great difference. Not one out of every hundred put in is ever opened for any purpose. In kitchen sink wastes, trap screws are very desirable, but in most small branch-waste traps they are rather more ornamental than useful.

Limited utility.

Desirable in sink wastes.

CHAPTER V.

WATER-CLOSETS.

The water-closet, though commonly supposed to be one of the "modern conveniences," has come to us from a very early and primitive civilization. Ewbank—the best authority on all matters pertaining to water and its uses—says they are of very ancient and probably Asiatic origin. They were introduced into Great Britain during the reign of Queen Elizabeth, but long before that time they were known, and to some extent used, in France and Spain. Their general employment is, however, of comparatively recent date, and from being a luxury of the rich, they have, within less than half a generation in this country, become the conveniences of nearly all classes of the population of sewer-drained cities. In country districts their use is limited, except in houses of the better class owned or occupied by people accustomed to the conveniences of the town.

Antiquity of water-closets.

Their general employment of recent date.

In itself considered, the water-closet is not of necessity an objectionable fixture in a sewer-drained house. Much as it has been inveighed against, it has many practical advantages over any other form of closet or privy. The question whether it is possible to dispose of the sewage of a town by other and better means than by water carriage, is not one which need be discussed here. Water carriage is not without objections, but it is likely to be generally employed in this country for many years to come, and we may, therefore, accept it as one of the conditions with which we have to deal. With water carriage the water-closet seems to have no practicable substitute. The individual householder may have recourse to earth closets, ash closets, absorbent buckets or any of the other dry conservancy systems, each of which has certain advantages, but

Water-closets not necessarily objectionable.

Water carriage.

Substitute for water-closets.

unless such appliances used receive more careful and intelligent attention than they are likely to get at the hands of servants, they will be found less satisfactory than the water-closet.

Privies. Privies are still used to some extent in cities, but there is certainly nothing to be said in favor of them. They are, I think, nuisances which should be abated by law, and if their abolition cannot be effected, they should be allowed to exist only in modified form and under strict sanitary inspection. The only objections to earth closets in cities and towns are the trouble and *Dry conservancy systems.* expense they entail. All the dry conservancy systems are open to the same objections, unless their employment is so general that some system can be organized for disposing of the contents of the receiving vessels without much expense, if any, to the individual householder. When each person who adopts one of these systems must pay for materials on the one hand and labor and cartage on the other, he will be very apt to conclude that the trouble and expense are far greater than the resulting benefit. Probably the time is not far distant when we shall realize more fully than now that we cannot afford to waste the wealth we pour into our polluted rivers, and which is needed for the enrichment of farm lands exhausted by what Baron Liebig forcibly characterized as "vampire agriculture."

Victor Hugo on the waste of Paris. Victor Hugo draws the following startling picture of the sewage waste of a great city, in "Les Misérables:"

"Paris throws five millions a year into the sea, and this without metaphor. How and in what manner? Day and night. With what object? None. With what thought? Without thinking. For what return? For nothing. By means of what organ? By means of its intestine—its sewer.

Chinese economy. "Thanks to human fertilization, the earth in China is still as young as in the days of Abraham. Chinese wheat yields a hundred and twenty fold. There is no guano comparable in fertilizing power to the detritus of a capital. * * *

Value of sewage. "We fit out convoys of ships at great expense to gather up at the South Pole the droppings of petrels and penguins, and the incalculable element of wealth which we have under our hand we send to the sea. All the human and animal manure which the world loses, restored to the land instead of being thrown into the water, would suffice to nourish the world.

"These heaps of garbage at the corners of the stone blocks; these tumbrils of mire jolting through the streets at night; these horrid scavengers' carts; these fetid streams of subterranean slime which the pavement hides from you—do you know what all this is? It is the flowering meadow; it is the green grass; it is marjoram and thyme and sage; it is game; it is cattle; it is the satisfied low of huge oxen at evening; it is perfumed hay; it is golden corn; it is bread on your table; it is warm blood in your veins; it is health; it is joy; it is life. Thus wills that mysterious creation which is transformation upon earth and transfiguration in heaven. * * * What sewage represents.

"Imitate Paris, you will ruin yourself. Moreover, particularly in this immemorial and senseless waste, Paris herself imitates. These surprising absurdities are not new; there is no young folly in this. The ancients acted like the moderns. 'The Cloacæ of Rome,' says Liebig, 'absorbed all the well-being of the Roman peasant.' When the Campagna of Rome was ruined by the Roman sewer, Rome exhausted Italy; and when she had put Italy into her Cloaca, she poured Sicily in, then Sardinia, then Africa. The sewer of Rome engulfed the world. This Cloaca offered its maw to the city and to the globe, *Urbi et Orbi*. Eternal city, unfathomable sewer. In these things, as well as in others, Rome sets the example. This example Paris follows with all the stupidity peculiar to cities of genius." The Roman cloaca.

This powerful description of a city's waste is certainly not overdrawn. No doubt Paris does throw five millions a year into the Seine, but the simple statement of this fact might lead to mistaken conclusions respecting the commercial value of sewage. The popular ideas on this subject, which find frequent expression in newspaper articles on sewage utilization, are The commercial value of sewage.

Sewage of English cities. clearly erroneous. The commercial value of the sewage of English cities, taken as it runs, is estimated to be about four cents per ton. Under favorable conditions, sewage irrigation can be carried on successfully, but most of the experiments in this direction have been unsuccessful, and very few of the large and costly sewage farms in Great Britain have paid or are likely **Sewage of American cities.** to pay interest. The value of Boston sewage is estimated by the Massachusetts State Board of Health to be about one cent per ton. The value of New York sewage is still less per ton, owing to its more liberal and even wasteful use of water per head **Sewage utilization.** of population. Practically it may be said to have no value, notwithstanding the fact that it contains a vast amount of valuable fertilizing material. With us the problem of the economical and profitable utilization of sewage will probably remain unsolved so long as we continue to dilute the valuable constituents of sewage with unlimited quantities of water. Great as its aggregate value may be, it is not usually great enough to pay for the separation of that which is valuable from that which is not. When necessity shall compel us to save this wealth, we shall probably begin by keeping that which is valuable out of the sewers altogether, and using these conduits only as runways **Water-closets not likely to be soon superseded.** for dirty water and street washings. Then we shall probably follow some system of dry conservancy, but for the present the water-closet remains an established institution, and it is useless to condemn it as unsanitary. The charge is well founded in a majority of cases, but vastly more of good will be accomplished by an intelligent effort to improve the water-closets now in use and remedy their defects than by riding Quixotic tilts against them. In my experience I have never found any good reason **Water-closet nuisances.** why water-closets should be either unwholesome or offensive. It is quite certain, however, that when any defects exist in the drainage system of a house, they will commonly be found conspicuous in the arrangements pertaining to the water-closets, and the sanitary inspector is not likely to make a mistake if, when called upon to correct the evils pertaining to bad plumbing work, he begins there.

WATER-CLOSETS.

The requirements of a good water-closet are various and imperative. To be safe and sanitary, as well as convenient, it must be inodorous. If it has any odor at all it will of necessity be a foul one, disgusting to the sense of smell, nauseous to the stomach and depressing in its general effect upon mind and body. If not actively poisonous, it will probably lower the physical tone of all who are compelled to breathe the air thus vitiated and predispose them to sickness. Often, however, it will be found actively poisonous, even when the smell, in itself considered, is not strong enough to noses accustomed to it to excite alarm or occasion inconvenience. *The requirements of a good water-closet.* *Offensive smell.*

It must be so constructed that it can be effectually flushed or washed out without waste of water; it must be so simple and strong as not to be disordered by average usage; and it must not afford an outlet at any point for the gases of the sewer. Obviously, but few water-closets in common use conform to these essential conditions. Nearly all will do so when new, if properly put up, but very few, comparatively, can stand the test of use, even for a brief period, without becoming nuisances. There is, therefore, much room for reform in water-closets, even though there exists no hygienic necessity for the abandonment of the principle of water carriage—regarding the question from the standpoint of the individual owner or occupant of a sewer-drained house. *Flushing.* *Strength.* *Common defects.*

To discuss fully and impartially the merits and defects of all the water-closets in general use in this country, would be a delicate and difficult task. Every manufacturer is jealous of his particular product, and every inventor stands ready to champion to the death his own particular invention and improvement. There is nothing in this menace of endless and bitter disputation to deter the conscientious writer from expressing his honest convictions; but in the case of water-closets this particularity is unnecessary as well as undesirable. They can be accurately divided into certain general classifications, and as the defects of a given closet are common, generally *Classification* speaking, to the class to which it belongs, a few remarks will serve to assist the reader in judging of the several styles competing for general favor, and in determining whether the ones he uses or finds in use are safe or unsafe.

Pan closets. Of the water-closets in use in this country, a very large proportion are pan closets. Closets of this pattern are defective in principle and unsatisfactory in operation; and although they have been variously modified and improved during the past few years, it is doubtful if they are susceptible of such improvement as will wholly correct their inherent defects. *Construction.* In Fig. 12

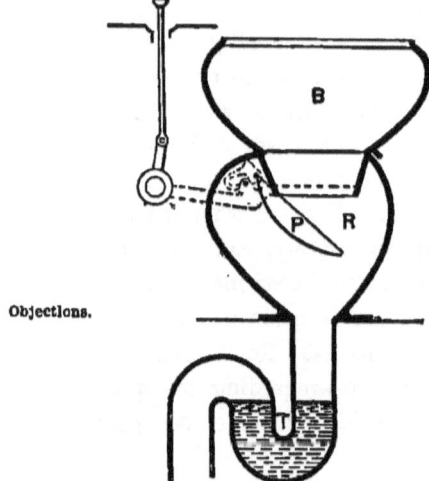

Fig. 12.

is shown a sectional view of the common pan closet. B is the basin, usually made of earthenware. P is the pan which, when the handle is raised, is tilted, emptying its contents into the receiver or containing vessel R, from which they pass into the soil-pipe branch, having a trap at T. *Objections.* As commonly made, the principal objection to this form of closet is that in addition to allowing the foul air from soil pipe and sewer to escape, it increases the mischief by manufacturing a great deal of very offensive and dangerous gas on its *Foul receivers.* own account. In use, the side of the receiver against which the pan discharges its contents when tilted, becomes coated with a mass of foulness that clings to and cakes upon it. Plumbers are frequently—though not so often as they should be—called upon to remove the receivers of pan closets and burn them out in order to abate the stench which has become intolerable. This affords only temporary relief, for the process of accumulation begins again at once and continues until the pan has barely room in which to swing. It is no uncommon thing to find the inside of a receiver so clogged that it is difficult to empty the pan. Matter thrown out as it descends is often scooped up again as it returns to its place. Human excrement is naturally viscous and plastic, and when dashed against a rough surface, like the inside of an iron casting, it takes hold and clings, especially as it is untouched by water which rushes down the basin and into the soil pipe when the pan is tilted. From this cling- *Poisonous gases from pan closets.* ing matter undergoing decomposition—more or less rapid according to circumstances—there constantly arise gases as poisonous as any which come from the foulest of sewers. These gases are held under constantly increasing pressure between the two seals, and when the upper seal is broken by tilting the pan, the water thrown down must, of course, displace the gases and force them out. If not thus liberated they will come *Transmission of gases through seals.* through the water seal in the pan, as explained in the preceding chapter. If the trap in the soil-pipe branch under such a closet is unventilated, its seal offers very little effective resistance to the upward passage of gases from the sewer and pipe. As the *Insufficient flush.* rule, the flow of water into the closets of this description is insufficient to flush the trap; consequently, the seal is seldom completely changed, and there is almost always a mass of undissolved fæces floating therein. With all these facts in mind it is easy to understand why a disagreeable smell, and sometimes an overpowering stench, is instantly emitted when the pan seal of such a closet is broken.

To obviate these serious defects in the pan closet, many de- *Modifications and improvements.* vices have been proposed, some of which are not without great advantage in modifying its worst evils. It is rather singular that although the nuisance and danger of foul receivers have been recognized for years, until recently no one seems to have thought of the simple experiment of flushing them. In a closet *Flushing receivers.* lately brought to my notice the principle of contraction was substantially as shown in Fig. 13. A represents the receiver and B the bowl, both of the ordinary form used in the common pan closet; C is the pan. The bowl receives water at the top from the pipe D in the ordinary manner, both valve and levers being operated by the handle as usual. Around the

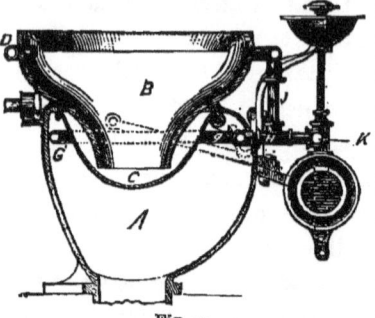

Fig. 13.

interior of the receiver A the pipe G is arranged in the form of a ring. This pipe is attached to and receives water from a pipe, H, which passes through the receiver A and communicates with the pipe D through a conducting pipe, J, so as to be supplied with water by the operation of the valve at the same time that the pan C is supplied. The pipe G is provided with perforations, g, through which the water escapes in streams or jets. The pipe G is also connected with a pipe, K, which communicates directly with the water supply pipe, and is provided with a stop cock which may be turned by hand in order to supply the pipe G with water independently of the pipe D and the valve. When in operation, the water issuing from the pipe G washes the inside of the receiver and the outside of the pan and lower part of the bowl, thus keeping the parts clean and preventing the outside of the pan from becoming corroded from ammonia. If, as frequently happens when the pan is emptied, the lower trap is syphoned, the flow of water from the pipe G overcomes this difficulty and promptly furnishes a supply for the trap. In emptying the pan the flow of water upon the inside of the receiver prevents the adhesion of soil and washes away the contents of the pan. This improvement is very valuable one, and will do much to abate the nuisance of foul receivers.

Venting receivers. Venting the receiver is an old expedient, and a good one in itself considered, but I do not believe that any amount of venting will make the common pan closet safe and excellent. Good ventilation is quite a different thing from simple venting. Ventilation implies an inflow as well as an outflow, and a natural or induced current. It is possible to secure these essentials if the proper means are taken, but such ventilation as is secured by carrying an air pipe into the side of the receiver does not usually suffice, although it is a great deal better than nothing.

Of late some currency has been given to the idea that it is possible to correct the defects of the pan closet by disinfection, and an apparatus to apply the disinfecting fluid has been invented and introduced in England. The chloralum or other disinfecting material is held in an earthen vessel above the closet, and when the valve is opened a small quantity is automatically let down through a pipe discharging into the basin, to mingle with the water of the flush and pass with it through the soil pipe to the sewer. The device has very little practical utility, if any. We have no disinfecting material cheap enough for free and general use which is sufficiently powerful to render any important service when thus employed. A good water-closet, properly set, amply flushed, kept clean and discharged into soil pipes open at both ends, will not need disinfecting; others will scarcely be benefited by it. *Disinfection.*

Of closets embodying the essential principle of the pan closet with certain modifications, there are a great many, few of which have come into general use in this country. Of these the valve closet (Fig. 14) is probably the most generally used. Some of these are a little better than the average pan closet; others have all its defects besides some which are peculiar to themselves. Few of them will hold their basin seal for any length of time when worn with use. Nearly all of them will syphon out this seal through the overflow upon very slight provocation. They are mostly inex- *Valve closets..*

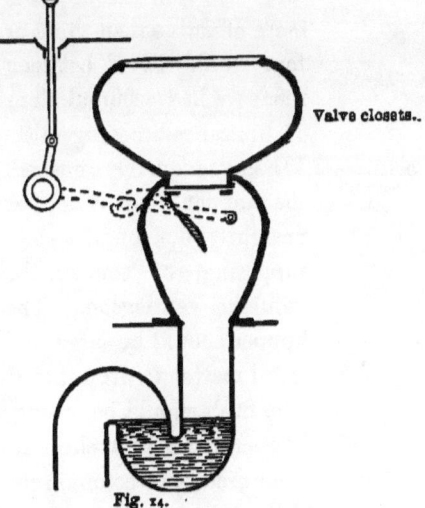

Fig. 14.

pensive affairs made for cheap work, and are not usually recommended for use in well-drained houses, even by those who make them.

Hopper closets.

Advantages over other cheap apparatus.

Form of hopper.

Plain hoppers, without either pans or valves and requiring no overflows, are in many respects the best of the cheap water-closets. Their advantage consists in strength, simplicity and the impossibility of concealing filth within them. They will become foul and offensive if neglected, but the fact that they need attention is apparent to the sense of sight, whereas the closets last described secrete filth within while clean on all visible surfaces. A neat housekeeper would not allow the sides of a hopper to remain foul, but the inside of a pan-closet receiver is beyond her reach and out of her sight. The usual form of hopper closet with automatic valve flushing apparatus is shown in Fig. 15. In closets of this kind we have but one seal — that in the trap made in the soil-pipe branch—but when this one seal is relieved from pressure by ventilation, it is more effective than three or four would be if between them we had accumulations of nuisance-breeding filth.

Conditions of security and efficiency.

When adequately flushed, the hopper closet is neither unsightly nor unsafe—presupposing, of course, the requisite ventilation. The hopper should be large and of such shape that it will permit fæcal matter to drop into the trap without touching the sides. The flush should be strong enough to scour the inside of the hopper and give a clean seal in the trap after the closet has been used. In common plumbing work the apparatus is usually

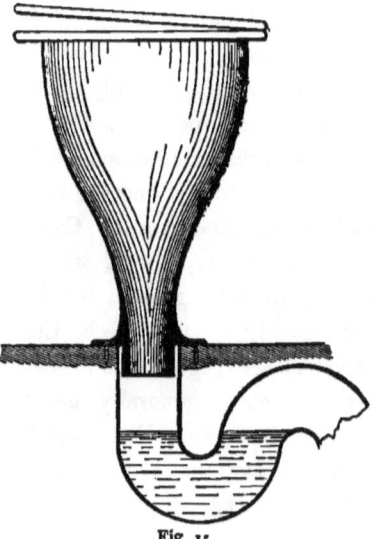

Fig. 15.

WATER-CLOSETS. 95

so arranged that a pressure upon the seat opens a valve and *Automatic flush.* allows a small quantity of water to trickle in as long as the closet is in use. Such a flush is of very little value beyond wetting the sides of the hopper and preventing accumulations upon them, and needs to be supplemented by a strong dash of water, the volume of which is under control of the person using the closet. When this is provided, either alone or, preferably, *Advantages of a double flush.* in connection with a moderate flow maintained automatically by means of a valve opened by a downward pressure upon the seat and closing when that pressure is relieved, the hopper closet with ventilated connections with the sewer should, if kept clean, be the best cheap closet in use.

Though seldom introduced into houses, the device known as *Ship closets.* the ship closet merits attention as a very perfect and sanitary apparatus. One of the best forms of this variety of water-closets which has come to my notice is shown in Fig. 16. The *Construction.* hopper is of good shape, but instead of being connected at the bottom with the soil pipe in the usual manner, an iron pipe makes a connection with the pump barrel seen on the left, the plunger of which is shown raised in the drawing. When the *Action.* handle is lifted, as shown, the contents of the hopper are drawn into the body of the pump, and at the same time a quantity of water is forced from the upper part of the pump barrel into the hopper through the pipe which connects them near the top. When the handle is depressed the soil, &c., contained in the body of the pump is forced out through a valve in the soil pipe rising behind the pump barrel, being prevented from returning to the hopper by a valve in the bottom of the pump. As the handle is forced down a vacuum is formed above the plunger, and through a check valve in the pipe on the left a supply of water is drawn into the body of the pump. The plunger is made large in order to reduce the quantity of water needed to fill the body of the pump at each stroke. The flush of water in the hopper on lifting the pump handle is forcible and abundant. As these closets are intended to be used below the water line on shipboard, it is necessary that they should be water tight in every part, as any inflow through them would

Fig. 16.

be a dangerous leak in the vessel. They are, therefore, so made that they can withstand a strong backward pressure in the soil pipe without difficulty. As made for use at sea, this form of closet is mostly brass, substantially put together and so arranged that it can be readily taken apart. Its positive movements and efficiency in preventing any back flow through the soil pipe, render it a very perfect device

Modification of ship closet for houses. regarded from a sanitary point of view. A modification of this form of closet, simplified, cheapened and adapted to use in houses, has been made but is not yet in use.

In buying a water-closet, it is a safe rule to buy the best that can be obtained and have it put in by the most intelligent and **The best closets the cheapest.** skillful plumber whose services can be secured. This is not only the best policy from a sanitary point of view, but it is the most economical in the end. A cheap, flimsy closet cannot be expected to stand the test of long usage, and will either be constantly out of order or must receive constant repairs. The essential requirements of a good water-closet have already been briefly stated at the beginning of this chapter, but with a view to fixing them in the mind of the unprofessional reader, I will now consider them more fully. They are:

Water losets must be free from offensive smell. 1. *It should be inodorous.* This is a quality of excellence which few cheap water-closets possess when in use. A faint, depressing, characteristic odor is almost always noticeable at or near the seat level, and very commonly in all parts of the room in which a water-closet is placed. Sometimes it permeates the air of the whole house. This is less readily detected by one accustomed to it from the impression it makes upon the sense of smell than from its depressing, headache-inducing effects. Those accustomed to the characteristic smells of city houses seldom notice this odor, except for a moment when coming in out of the fresh air; but it has often been remarked by people living in the purer atmosphere of the country that they notice the smell of the sewer the moment they enter a city house. I have no doubt this is true, for although I have lived the greater part of my life in cities, I find more difficulty in escaping than in detecting the smell of the sewers when passing through most city houses. The water-closets are not wholly to blame for this, but they are largely so. For this reason, the rooms in which water-closets are placed should be well ventilated all the time. The too general custom of placing them in little pantries, with no ventilation except through a shaft opening into similar pantries above, and lighted by a water-tight glass skylight, is bad. When not ventilated at all, as is often the case, the evil is still greater. Ventilation at the level of the seat is recommended by some plumbers, and when secured it is doubtless good. Closets so constructed, in connection with ventilating shafts, as to provide for a constant current down through the basin and out, have been introduced and are well spoken of. I have had no personal experience with them, but if the ventilation for which provision is made can be secured, I can see no reason why they should not work well. But whatever the form of closet used, the architect should see to it that the apartments in which they are placed are well ventilated throughout.

Characteristic smell of city houses.
Rooms containing w. c. should be ventilated.
Misplaced closets.
Ventilation.

While upon this subject I may remark that a great many people, especially young women, pass much more time than is necessary in the usually unwholesome atmosphere of these places, taking books and newspapers with them. This is a vicious habit which entails more than one evil consequence upon those who practice it; but it is nevertheless very common. People should learn, and children should be taught, to regard a water-closet as just what it is. It certainly is not a place in which to pass an idle hour reading novels.

Water-closets not reading rooms.

98   WATER-CLOSETS.

Examples of bad water-closets. Two instances out of the many which have come to my notice will serve to show to what extent sanitary principles are disregarded in much of the "bread-and-butter architecture" of the present time. Both were found in fine brown-stone houses, commanding a high rental and occupied by physicians of excellent professional standing. In one of these houses I found pan closets of the worst description, placed in little pantries with a floor space only equal to about twice the area of the seat. These pantries had absolutely no light except that provided by gas burners, and no ventilation except what was had through the doors opening into the halls, and through little windows scarcely larger than a sheet of letter paper, opening into adjoining apartments. The closets were neglected and almost always offensive, but they had been tolerated for years without attracting attention, notwithstanding the fact that during this period sickness was the rule rather than the exception in the family. In the other case the arrangement was very much like that just described except that the pantries were rather more *Ventilating water-closets into halls and bedrooms.* roomy. They were, however, without ventilation except such as might be secured through the halls or the adjoining rooms. The smell, noticeable at all times, was often sickening, but I was unable to convince the physician that there was any objection to the arrangement great enough to justify him in ventilating his soil pipe and abating the nuisance to which the water-*Precept vs. practice.* closets continually gave rise. It is a curious fact, but one which has often been remarked, that those who are most ready to discover and inveigh against unsanitary conditions in other people's houses, are the most difficult to convince that their own houses are not so well arranged as they should be, even though conspicuously unhealthy. It is probably for this reason that sanitarians, as a rule, live in houses that will not bear inspection.

Importance of an abundant supply of water for flushing. 2. *It must be abundantly and frequently flushed.* This is easily provided for. Most closets are so constructed that they can be flushed abundantly if the handle is held raised long enough; but the common practice is to give the handle a jerk and let it drop. For this reason it is important that the flushing apparatus should be so arranged that a strong and voluminous flow is secured, even though the handle is raised only for an instant. This is provided for in most of the valves in use in good plumbing practice, by an arrangement which permits the valve to close very slowly when once opened. If for any cause the flow of water into and through a closet is not enough to scour it out thoroughly and leave a clean seal when the flow ceases, a competent plumber should be called in to remedy the defect, which he can commonly do in a few minutes. *Flush valves.*

Water for flushing closets should not be drawn from a service pipe in which the flow is liable to be interrupted by the opening of a cock between the closet and the main. To this rule there are no exceptions. Unless each water-closet has a separate service pipe which is not tapped for any other purpose, some form of cistern or its equivalent should be provided and the water needed for flushing drawn therefrom. In cheap contract plumbing work, however, it is not unusual to find the main service pipe tapped on every floor for all purposes, water-closet flushing included. Where the Holly system is used, the pressure of water in the mains is usually so great that a service pipe might be tapped several times without interrupting the flow at the highest cock or valve. In New York, however, and many other cities where the pressure in mains and service pipes depends upon the head of water in distributing reservoirs, it is the rule rather than the exception to have the supply cut off from the upper part of a house by the opening of the faucets in the kitchen. This can be avoided, but it seldom is except in the best work. Under these circumstances, water cannot be drawn direct from the service pipe into the basin of a water-closet without serious danger. Cold or cool water, especially if pure, has a vast capacity for absorbing gases. Ordinarily it will absorb many times its own volume, and all sorts of gaseous impurities are taken up by it as readily as a dry sponge takes *Service pipes should not be tapped direct into water-closet basins.* *An evil characteristic of cheap work.* *Absorption of gases by water.*

up water. Such impregnation of course renders water foul and wholly unfit for drinking or use in culinary operations. It is for this reason that the service pipe of a house should not be connected with water-closet basins when, from any cause, the flow is liable to be intermittent.

Evils of tapping service pipes into water-closets. Suppose that when water is drawn in the kitchen and the flow in the upper part of the house is temporarily interrupted, the handle of a water-closet is raised when the basin is foul. The water in the pipe, which might otherwise have been held there by atmospheric pressure, runs down as soon as air is admitted at the top, and the current of air which rushes in to fill the pipe carries with it the impure gases from the water-closet basin. *Water pot soning.* These are at once absorbed by the moisture clinging to the sides of the pipe, and water subsequently drawn from this pipe may be very impure and unwholesome. A striking illustration of water poisoning by the gases generated in water-closets is found in the sanitary history of the town of Lewes, *Typhoid fever at Lewes.* England. During the year 1874 typhus fever broke out within the corporate limits, and nearly five hundred cases were reported within a few days. The town is supplied with water by three private companies, much as our American cities are supplied with gas. The fever cases occurred in the districts *Intermittent water service.* supplied by the Lewes Water Works Company. The supply furnished by this company was not constant, but was turned on for three or four hours in the morning, and again during the afternoon. When the water was shut off the pipes emptied themselves, and air rushed in at every opening to fill the partial *Air from closet basins drawn into water pipes.* vacuum created by the outflow of the water. During the prevalence of the fever an investigation was made, and the following facts were brought to light: Many of the water-closets in the infected district were flushed with water drawn directly from the service pipes, and it was the habit of the people to have the valves open in case these closets were used when the water was shut off, so that they might be flushed as soon as the flow was resumed. The consequence was that a part of the air which entered the mains was drawn in through the basins of foul closets. Other defects were found of a still more serious nature, by which the gases of foul privy vaults were drawn into the service pipes through cracks and holes made by corrosion of the lead. Here was found the cause of the outbreak of fever. In one district sixty houses out of four hundred and fifty-four were supplied from the works of the Lewes Water Company, and in this district the cases of typhus fever occurred only in these sixty houses, with the exception of two cases. The other three hundred and ninety-four houses were free from it. Even after the epidemic had become general, and was propagated by other means, only six per cent. of the total number of cases occurred outside of the houses supplied with this company's water. *Extent of the epidemic.*

Other illustrations might be given of the danger of tapping sewer pipes direct into water-closet basins, but the fact that it is dangerous will be evident to every intelligent person without further argument. I am aware that plumbers are often called upon to do all sorts of objectionable things, and that an attempt on their part to explain the necessity for employing better methods may be regarded as an effort to sell more material and make a larger job to charge for. But while I am not prepared to say that a conscientious plumber should refuse to contract for such work, I have no hesitation in saying that he should not do so without explaining the danger of mistaken economy, thus clearing himself of moral responsibility for the consequences if his protest is disregarded. It is not my purpose to discuss the ethics of this question in this chapter, if at all, but it seems to me that the Health Boards of New York and other American cities should endeavor to secure such additions to the local building laws as would relieve plumbers of the moral responsibility of doing bad work and justifying it to their consciences with the specious plea advanced by the starved apothecary of Mantua : *The moral responsibility for bad plumbing work.* *Plumbing laws needed.*

"My poverty, but not my will, consents."

3. *It must be strong, simple and not liable to get out of order.* This is a qualification demanded for sanitary as well as *Strength, simplicity and durability.*

for economic reasons. A complex water-closet, weak in any part or liable to derangement from any cause, is quite certain to be a nuisance sooner or later. When water-closets were first introduced into the American market, they were made large, strong and substantial. Of late years there has been a tendency to diminish their size, lighten their parts and cheapen their cost. As the result, most of the water-closets now in the market are weak, flimsy affairs, not worth the low price at which they are sold. There is no economy in buying cheap water-closets, whatever the character of the houses in which they are to be placed.

Deterioration of water-closets.

No economy in cheap goods.

4. *It must be sealed against the inflow of air currents from the sewer and soil pipe.* This cannot be secured with certainty unless the ventilation described in the chapter on traps and seals is provided. It is simply impossible to maintain a clean and efficient water seal in any trap discharging into an unventilated sewer connection. As already shown in these pages, water offers no effective resistance to the passage of impure air or light gases, and unless these are afforded an easy and direct outlet to the open air from the sewers and drains in which they form, they will not be held back so long as there is nothing to prevent their escape except a small quantity of water which will eagerly absorb and as readily transmit them. From a somewhat careful examination of the various patterns of water-closets in use in this country, I am forced to the conclusion that but few are to be recommended. So far as I can judge, I should say that the best are those which command the highest price and are least often called for in architects' specifications. A water-closet which combines the requisite qualities above specified has never yet been made at a price which will admit of its employment in cheap work, and probably never will be. The problem of constructing a closet which shall be at once cheap and good presents many difficulties, as any one will see who, with an intelligent idea of what there is and what is needed, attempts its solution.

Effectual sealing.

Impossibility of sealing unventilated pipes with water.

Few water-closets to be recommended.

Relation of quality to cost.

Difficulty of combining cheapness and excellence.

WATER-CLOSETS.

The only article in plumbing work akin to the water-closet, **Slop hoppers.** and which can properly be considered under the same head, is the slop hopper. This is a cone-shaped hopper of iron or earthenware, connected with the soil pipe by a trapped branch. Slop hoppers are usually placed on the upper floors of large houses, hotels, &c., to afford a means of disposing of waste water without throwing it into the water-closets. For this they **Utility.** are useful, and should always be provided in large houses. As commonly put in, however, they give rise to no little mischief. Water is usually poured into them by the bucketful, and the **Unsealing traps.** strong and sudden downward flow thus provided is very apt to suck the seals out of the traps in some of the small branch wastes. This could be obviated in most cases by ventilation, **Strainers to retard the** but additional security is afforded by providing slop hoppers **overflow.** with strainers so fine that the water thrown into them will not pass out any more rapidly than the soil pipe can carry it off without danger to the seals in traps. When traps are not provided with independent ventilation, I consider this precaution indispensable to safety. Servants cannot be expected to exercise any discretion in such matters.

When from any cause the water supply of an inhabited, sewer-drained house is cut off, the water-closets demand immediate attention. At such times they usually become very foul, and from negligence in renewing water seals in traps, pernicious exhalations from the sewers vitiate the air of living and sleeping rooms. The accumulation of excremental matter in basins, receivers and traps cannot but give rise to dangerous nuisance, which is little bettered when just enough water is thrown in to wash these accumulations out of sight and into the containing vessel or mouth of the soil-pipe branch.

CHAPTER VI.

SERVICE PIPES AND WATER SERVICE IN CITY HOUSES.

The occupant of a city house has, as the rule, no choice as to the character of the water he uses. His only available source of supply is the street main, and, generally speaking, he has no occasion to trouble himself in the matter further than to see that the service system of his house is so arranged as to *Constant ser-* secure, if possible, a constant supply at every cock. This *vice a prime requisite.* cannot always be had without pumping, even in cities abundantly supplied with good water. In many parts of New *Tanks.* York, for example, it is necessary to pump into tanks or reservoirs, near the roof, all the water drawn above the basement floor. In this case the service pipes are carried down through the house, which involves a somewhat different arrangement of them than is needed when the supply rises from pressure in the mains.

Pressure When the pressure of water in the mains is sufficient to *due to head.* insure a constant service on every floor, as in many cities, especially those in which a head is secured by means of powerful pumping engines, there is no occasion for any mistakes on the part of the plumber who understands the mechanical part of his trade—so far, at least, as securing a good distribution is *Arrangement* concerned. The plan upon which the pipes are arranged must, *of pipes.* of course, depend upon the size, character and internal arrangement of the house. The architect should, and sometimes does, make suitable provision for water service and drainage in his plans, but too often he designs the house in every detail first and then provides for the pipes in whatever way he finds easiest.

Conditions of The conditions of a good water service are that it shall be *a good water service.* constant, in well-laid pipes protected from all danger of freezing, bursting or leaking from any cause, and so placed that,

SERVICE PIPES AND WATER SERVICE IN CITY HOUSES. 105

when the supply is shut off, the whole system may be drained by opening a waste cock in the cellar or kitchen. In the chapter on waste and soil pipes, I ventured some suggestions respecting the best disposition to be made of the pipe system in a house. Further on in this chapter I shall speak of the means which I have found most efficacious in protecting service pipes from frost at the points where they are most exposed.

The question of the best material for service pipes is one of great importance, and it has been long under discussion without acquiring much interest for the general public. If the question of health were not involved, the problem would be a very simple one, for lead pipe, taken for all in all, would probably be the one universally preferred. Lead, from its cheapness and the facility with which it can be manipulated cold, has been for centuries the favorite metal of plumbers, and the name of the trade is derived from the Latin name of the metal. With the exception of ease of working, however, lead cannot be said to possess qualities which adapt it for use as a material for service pipes. Lead pipe is heavy and weak; it readily stretches, and sags or buckles when exposed to variations of temperature; it is easily crushed; rats can cut it without difficulty, as they sometimes do, and many kinds of water attack, corrode and are poisoned by it. For the reason last mentioned, lead has long been regarded by chemists as an unsafe metal for service pipes. Observation and experiment seem to have confirmed this opinion; and that, under a great variety of conditions, it will be found to rest upon a substantial basis of scientific truth, I shall attempt to show in Chapter VIII.

In the effort to find something better than lead as a material for service pipes, experiments have been made with nearly all the cheaper metals, as well as with a good many substances not metallic, such as glass, paper, gutta percha, &c. Block tin pipes have been extensively used, but they are not, and probably never will be, regarded with general favor. Tin is harder

Material for service pipes.

Lead pipes.

Objections to lead.

Regarded as unsafe.

Non-metallic substitutes for lead.

Tin pipes.

and much less ductile than lead, it melts at a lower temperature and, in addition to being a more costly metal, it is much less easily manipulated. In some kinds of water it is rapidly corroded, and my observation leads me to believe that those waters which act upon tin most rapidly do not attack lead vigorously, if at all. I may be mistaken on this point, but I have frequently had my attention called by plumbers in country districts to instances of the rapid destruction of tin pipe in waters which apparently had no effect upon lead; while in as many other cases I have found tin to render excellent service under conditions fatal to the life of lead pipe. Fortunately, the soluble salts of tin are not poisonous. Its cost and the difficulty of manipulating it are not objections which need stand in the way of its more general use, but they always have and probably always will.

Conditions of corrosion.

Lead-incased tin pipe—or, as it is more commonly called, tin-lined lead pipe—has been extensively used during the past few years, and has generally given satisfaction. In some waters the tin lining lasts but a short time, but in a great majority of cases it will be found a safe and durable pipe. As now made, this pipe has a continuous tin lining throughout, of very even thickness; and, when properly put together with the appliances furnished, and according to the method prescribed by the manufacturers, it will deliver water free from poisonous metallic salts as long as the tin lining lasts and perhaps longer. If, however, this pipe is worked in the same manner as lead, any but a very skillful and exceptionally careful workman will be pretty sure to leave lead surfaces exposed, thus defeating the object of the tin lining. The manufacturers of lead-incased tin pipes supply tinned brass ferrules, thimbles and T connections, with which perfect joints can be made, insuring a continuous tin lining, but plumbers, as the rule, object to using these and insist on wiping the joints after the usual practice. I know of no reason for this unless it be that the wiped joints are more profitable, as they enable the plumber to use a pound and a

Tin-lined lead pipes.

Advantages.

Disadvantages.

Tinned ferrules.

Wiped joints.

SERVICE PIPES AND WATER SERVICE IN CITY HOUSES. 107

half of solder, worth from 14 to 15 cents a pound, and charge it on the bill as five pounds at 50 cents per pound. The objection to wiped joints in tin-lined pipe is that, in "getting a heat" preparatory to spreading the solder after pouring, the lead becomes so hot that the tin is likely to be fused and destroyed as a lining, but remains in the pipe to obstruct the waterway. Joints in tin-lined lead pipe can sometimes be wiped without destroying the tin lining, but I would not trust any plumber to do it for me. I am not prepared to say that bad joints will make a tin-lined pipe no better than one of lead, but it is quite evident that the theoretical excellence of the former over the latter depends upon the maintenance of a continuous tin lining, and this plumbers are not usually willing to give them.

Wrought iron has been employed as a material for small pipes for water distribution with good hygienic results, but not without some disadvantages when unprotected from rapid oxidation. The water coming through iron pipe is not considered injurious, even when a considerable amount of the metal is present. The low price of wrought iron pipes, their great strength and the ease with which they are put up render them very desirable, provided they can be made durable by protection against rust. I have known of several instances in which wrought-iron tubes have been used as service pipes, but in only one case have I been able to get the facts necessary to the formation of an opinion as to their advantages compared with other pipes. This was in the case of Swarthmore College, at Swarthmore, Delaware county, Pennsylvania. Water was introduced into the college building through iron pipes in 1869, and from that time until March, 1875, at which date my inquiries were made, they had required and received no attention or repairs. In February, 1875, some changes were necessary, and a piece was cut out of one of the pipes which had been in constant use for six years. Its original diameter of bore—one inch and a half—had been reduced to about one inch by the accu-

Wrought-iron pipes.

Rust.

Iron pipes at Swarthmore College.

Results of six years' service.

108   SERVICE PIPES AND WATER SERVICE IN CITY HOUSES.

Analyses. mulation of rust on the interior surface. At my request the respected president of the institution, Prof. Edward H. Magill, caused a series of analyses to be made to determine the chem-
No organic impurities. ical effects of the iron upon the water. Samples were taken from the spring and drawn from the cocks inside the building, and the latter were found very pure, containing no trace of
Percentage of iron in water. organic matter. The most remarkable fact was that the percentage of iron in the water taken from the cocks was only about two-fifths of that found in the water taken directly from the spring. This was established by repeated determinations, and the fact cannot be doubted that the passage of the water through iron pipes caused it to deposit three-fifths of the iron it had previously held in solution. In its general character the water is considered excellent, and at the date of my information it was the opinion of the engineer of the college that the pipes were good for 15 or 20 years more. This is not as long a life as might be expected of lead pipe under favorable conditions, but when healthfulness is taken into consideration, there is really no comparison to be made between the two metals.

Galvanized iron pipes. Galvanized iron pipes have been in the market for many years, and when they were first introduced it was hoped and believed that they furnished a satisfactory solution of the prob-
Zinc unsafe. lem of a cheap, durable and safe service pipe. This expectation has been disappointed, and although still extensively used, pipes coated with zinc are regarded with disfavor by a majority of those
Imperfect protection to iron. whose opinions are entitled to consideration. In many cases the iron, unequally and imperfectly coated with the zinc, is more rapidly corroded and destroyed than it would have been if left unprotected. A great variety of waters used for the supply of towns attack zinc vigorously, and all the resulting zinc salts are
Zinc poisoning. poisonous. An effort has lately been made to show that the salts of zinc can safely be taken into the human system in larger quantities than could be held in solution in potable waters, but there are too many well-attested cases of zinc poisoning from drinking water conveyed through galvanized iron pipes to render the experiment a safe one.

SERVICE PIPES AND WATER SERVICE IN CITY HOUSES. 109

Many efforts have been made to tin iron pipes, but I have never heard that they were successful. The outside of the pipe could be very well coated when desired, but the porous and uneven character of the inside surface causes the tin to be unequally and imperfectly deposited, leaving portions of the iron exposed to the water. In this case the objection is a commercial one. The pipe is too rapidly destroyed to be economical. More recently a process has been perfected for the manufacture of a tin-lined iron pipe, which seems to overcome this difficulty. It is a wrought-iron pipe with a continuous and perfect independent tin lining. The tin pipe is drawn by the usual means, and of such size that it can be slipped inside the iron pipe of which it is to form a part. Hydraulic pressure is then applied by means similar to those employed in testing pipes, and the ductile tin lining is expanded, conforming to all the inequalities on the inside of the iron pipe and being thus locked firmly in place. Tin-lined iron fittings are also supplied by the manufacturers of this excellent pipe. In making a joint a tinned ferrule with a sharp lip, slipped into the end of the pipe, is so arranged as to take a bearing all around upon the tin lining of the fitting, and thus, while making a perfect joint, presenting a continuous lining throughout. These pipes are made in 16-foot lengths of all the usual sizes. It is, in my judgment, the best pipe for water conveyance ever made. If the tin lining is destroyed by the corrosive action of water, there is no danger of metallic poisoning as in the case of lead and zinc, and we still have a strong, durable and safe iron pipe behind it. The tin is so thick, however, that there is no danger of its complete destruction except under unusual and peculiar circumstances. A special cutting tool is used when it is desired to preserve the tin lining so as to turn it over the end, but the operation is no more difficult and takes no more time than to cut ordinary gas pipe with the common tool.

During the past few years iron pipes with glass lining have come into use and have secured some degree of public favor in

Tinned iron pipes.

Iron pipes with tin lining.

Process of manufacture.

Ferrules.

Joints.

Safety.

Advantages.

Glass-lined iron pipes.

spite of a very determined opposition from the plumbing trade. This opposition is probably due to the fact that plumbers cannot put them in. They are very difficult to cut with common tools; the brittle lining will not bear rough or careless handling, and when a pipe tongs is applied to one end of a length to screw it into a coupling at the other, a workman unaccustomed to the work would probably shiver the glass lining before he had brought the pipe to a solid shoulder. These difficulties have been met by the manufacturers. If a plan of the plumbing work of a house, with measurements, is furnished them, they will fill an order for the required amount of pipe cut to the proper lengths and threaded at the ends to fit the glass-lined joints. If the plan and measurements are correct the pipes will go together without trouble. They will not bear a very violent torsion strain, as the lining is brittle, but if the mechanic who puts them in—preferably a gas and steam fitter—uses his pipe tongs with judgment and expends no unnecessary energy, he will find the setting up of a line of glass-lined pipe a safe and easy matter. Pipes of this kind should be well protected from frost, as they are likely to be troublesome if allowed to freeze. I am informed that water held in glass-lined pipes will bear exposure to a temperature 14° Fahr. below that which would freeze water held in lead pipes. I cannot say from personal experience whether this is so or not, but I have it on good authority and am quite prepared to believe it, as there is a film of plaster of Paris between the iron and the glass.

Of enameled iron pipes there are a great many. Some that I have seen were no better for the "enameling" they had received, which was nothing more than a shiny coat of baking varnish; others are evidently well protected against corrosion. The best pipe of this kind which has come to my notice is covered inside and out with a thick, elastic enamel which has resisted all the tests to which I have thus far subjected it. In other and more competent hands it has been subjected to various severe tests with acids and alkalies, to boiling in water—

pure and charged for experimental purposes with various salts —and to the action of gases. In all cases the results were satisfactory.

Of "composito" pipes, made of alloys of two or more metals, there are three or four in limited use. Those which I have seen and tested have nothing to recommend them. One was badly made of very impure and inferior lead, and another, to which my attention was attracted by the announcement that it was "the long-sought and much-needed sanitary service pipe, which will deliver pure water and never wear out," I found to be principally lead. *Composition pipes. Lead in disguise.*

Brass and copper tubes have, in some instances, been used as service pipes, chiefly in and about Boston. Two or more large firms are engaged in the manufacture of these pipes, which are principally used by engine builders for the conveyance of steam. They are washed with tin inside and out, inside only, or not at all, according to the fancy of the buyer. They are light, strong and very durable, but I should not consider them well adapted for the conveyance of water for domestic use. The coating of tin has no appreciable thickness, and would not long protect the brass and copper from corrosion, and the salts of these metals are poisonous. From recent inquiries I learn that they are going out of use in plumbing work, and that where best known they are regarded with least favor. *Brass and copper pipes.*

There are still other kinds of pipe in the market, but their employment for the conveyance of water is so limited that it is scarcely worth while to describe them.

From the foregoing it will be seen that as regards material there is a pretty wide range for selection. A careful examination of the facts presented in Chapter VIII will further assist the reader in forming a correct opinion as to the kind of pipe best adapted for use under given conditions. *Opportunity for judicious selection.*

In putting in a system of service pipes the plumber should observe certain general rules, which cannot be disregarded with safety. Primarily, he must see that the whole system is so

112 SERVICE PIPES AND WATER SERVICE IN CITY HOUSES.

Emptying pipes. arranged that when the supply is shut off every drop of water in them may be drained out. To be able to empty the pipes is a great advantage to the housekeeper, and if judiciously availed of, will often avert much damage to walls, carpets and furniture. Unless fully protected against freezing, the water *When desirable to empty.* of a house should always be shut off at night and the pipes drained in very cold weather. This is inconvenient and dangerous, I admit, but it is less so than a protracted stoppage of the water service from ice in the pipes, which gives rise to conditions vastly more unsafe, regarded from a sanitary standpoint, than any likely to result from a stoppage of the service at night.

Should be provided for in plans. To so pipe a house that it may be drained dry requires judgment and skill on the part of the plumber; in some instances which have come to my notice it would have been practically impossible had the plumber not received the intelligent coöperation of the architect. It would be difficult to give any rules for the piping of a house which would admit of general application, but a few suggestions on this point may be of use.

Branches should have continuous support. When tin, tin-lined or lead pipe is used, all branches from the vertical lines should be given a continuous support and such an inclination that they will drain into the main service pipe with which they connect. The continuous support, which is secured by laying them upon shelves, obviates the otherwise *Sags.* inevitable formation of running traps. The sagging of pipes results from expansion and contraction incident to changes of *Expansion and contraction.* temperature. When the metal warms from any cause, it expands. As it cools it contracts, but does not return to its original shape. Lead is heavy and soft, and the force with which it contracts is expended in stretching the metal, which requires less power than to lift back into line the part that has sagged. *Creeping of lead.* This phenomenon is forcibly illustrated in the "creeping" of lead sheets laid upon roofs with even a slight pitch, in Great Britain and some parts of Continental Europe where lead is used for roofing purposes. The expansion of the lead when

SERVICE PIPES AND WATER SERVICE IN CITY HOUSES. 113

warmed by the sun is in the direction of the eaves. When it Lead roofs.
contracts again the sheet does not come back to its original
position, but the upper edge is drawn down, pulling out, in
turn, the nails with which it is held or tearing loose from them.
This creeping goes on until the lead has slipped off the roof—
its progress being much like that of a measuring worm. When Supporting
iron or other stiff pipes are used, a continuous support is proba- iron pipes.
bly unnecessary, but the hooks by which they are held in place
should clasp it loosely at frequent intervals. Vertical lines of
lead, or lead-incased tin pipes, are supported by lead flanges,
technically known as "tacks," which are soldered fast to the Tacks.

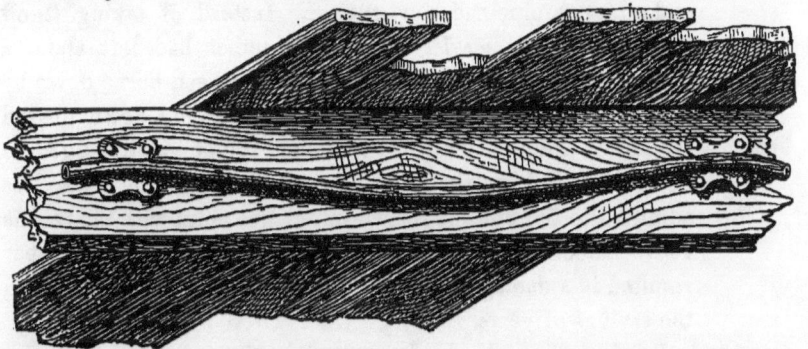

Fig. 17.

pipes and held to contiguous woodwork by screws. In good
work horizontal lines of lead pipe are provided with continuous Continuous
and substantial support, and in some of the best jobs I have safes.
seen, with continuous safes to catch and dispose of leakage.
When lead or tin pipes are laid under floors or in other positions nearly horizontal, with only such support as is secured by
occasional tacks, pipe hooks or loops of sheet metal nailed to
floor timbers, they are almost certain, from the causes mentioned, to form a succession of running traps, as shown slightly Running
exaggerated in Fig. 17. It may be accepted as a general fact traps
that any stretch in lead pipe, due to expansion or the stretching Stretch.
of the metal by its own weight, will be a permanent addition
to its length, and when running traps are formed, as shown in
Fig. 17, the pipes cannot be drained. Any defect in the
8

arrangement of the pipes which has this result may give rise to serious mischief. I saw an example of this a few years ago in one of the most elegant residences in New York. Owing to the absence of the family at Washington the house was not occupied during the winter, and the water was shut off and the pipes drained. In the spring, when the water was turned on, a leak was developed, and before it could be reached and stopped it had destroyed a superbly frescoed ceiling and side wall, ruined a valuable carpet and done injury to the value of many thousands of dollars. Investigation showed that at some time a change had been made in the pipe system of the house and a branch pipe had been cut off. Instead of taking it off close to the main service pipe, the plumber had left about a foot of the branch connected, and this had been borne down by its own weight until it hung suspended, pointing down toward the parlor ceiling. This had remained full of water, had frozen during the cold weather and burst, and when the water was admitted to the pipes it rushed out at this point with the results above stated. A little carelessness on the part of a plumber resulted in a damage greater than he could have paid for with the savings of years.

Importance of accessibility. Service pipes should be accessible throughout their entire length, so that a plumber can make any needed alterations or repairs without ripping up floors and making work for the carpenter. Plumbers are very good at this kind of work, and judging from the vigor with which they attack a floor or wall with whatever implements are most convenient, I should say that tearing up houses was congenial and agreeable occupation to many of them. Architects should guard against such destruction of woodwork and plaster by arranging the floor in such a way that a pipe can be reached without difficulty or delay. I would as soon think of trusting my watch with a blacksmith to put in a new mainspring, as give a plumber a *carte blanche* to operate on the woodwork of my house with a saw, hammer and cold chisel.

In piping a house with iron, the work belongs to the trade of gas and steam fitting rather than that of plumbing. These are now a part of the plumber's trade, however, and the expert is expected to know how to work on iron as well as lead. When iron pipes are used, no especial precautions are necessary except to see that every branch is given a sufficient descent from the point of discharge to the main service pipe. There is no trouble in securing this if the house is planned with intelligent reference to the proper arrangement of the pipes, but plumbers are often bothered by serious difficulties which the architect should have foreseen and provided for in his plans. It often happens, however, that the architect is compelled to modify his plans at the last minute to bring the plumbing of the house within the scope of human ingenuity. The lack of a practical knowledge in such matters, to which a great many architects must confess, accounts for much of the unsatisfactory work done by plumbers. *Piping with iron.* *Errors and omissions in plans.*

To so arrange the service-pipe system in a house that, so long as the average pressure is maintained in the street mains, or main supply pipe, we shall have a constant flow, if desired, at any and every cock and closet valve, is one of the refinements of the plumber's art which does not come within the knowledge of a majority of those who claim to know their trades. Certainly it is a desideratum for which comparatively few architects know how to provide without adding greatly to the cost of the plumbing work in a house. As a consequence, the service pipes in a majority of houses are so arranged that when a cock is opened in the kitchen or on the parlor floor, those seeking to draw water on the floor above must stand and wait patiently until the cock below is closed, the pipe refilled and the flow resumed. This often entails great inconvenience, and is sometimes attended with danger, as foul water-closets may be left unflushed and forgotten by those who, at the moment of using, cannot get a flush. When closets are flushed from tanks, which they should be under all circumstances, this dan- *The distribution of water through houses.* *Intermittent flow on upper floors.* *Unflushed water-closets.*

ger does not exist, but in the case of those flushed through a branch of the service pipe, the danger of neglect is, as before shown, further increased by the certainty that if the valve is opened when the flow is stopped in the service pipe, the column of water held up in the pipe by atmospheric pressure will at once drop, and foul air charged with organic impurities will be drawn in through the water-closet basin to fill the vacuum.

Inconveniences of a defective water service. But were there no danger to be apprehended, the inconvenience which those experience who live in houses piped in this slovenly and imperfect way, should lead to a reform. For the architect and the plumber no other consideration need have weight than that a house thus piped is, at best, a botched job. The water pipes can be so arranged that with but little added cost we can have a constant supply of water for every floor to which its head will carry it. The reader will understand, of course, that these remarks do not apply to houses in which water drawn above the kitchen floor is supplied from a tank to which it has been elevated by a force pump, but only to those so situated that a distribution is effected by reliance upon the pressure due to head in the mains.

Tapping a return pipe. The plan which first occurs to mind, and which was probably earliest adopted, is that of carrying the service pipe untapped to the highest point at which a service is desired, and connecting all cocks and branches with the return pipe. This plan can be adopted with safety only under exceptional conditions. *Conditions of successful operation* When we are sure of a head at the source of supply sufficient to raise the water, under all conditions, to the turn of the pipe, it will work very well. For example, in a country house supplied from a reservoir or constant spring situated on a hillside above the house, we could follow such a plan without danger of inconvenience. In cities, however, we are likely to have trouble with it, as the head varies, sometimes through a pretty wide range, and when it falls below the level of the bend in the pipe we are likely to have trouble. For the supply of cocks below the level of the head, the water will continue to syphon over the bend, but when, during this period of low pressure, a cock above the level of the head is opened by accident or design, the syphon is broken, the pipes empty themselves and the water is cut off from the house until the head rises. We must, therefore, have an air cock at the bend, and the trouble of giving this proper attention at the right moment, and the danger of having our water supply cut off at times when the head would be great enough to maintain a service on two or three floors, render this plan undesirable for adoption in city houses. It also entails the expense of a double service pipe, and renders difficult and troublesome the often necessary expedient of emptying the pipes when danger from frost is apprehended. The emptying is easy enough, but the refilling and reëstablishment of a flow through the pipe often involves some trouble. For these reasons the plan cannot be recommended for general use. *Air cock.*

Another plan, which gives each floor a separate service pipe, is free from objection save that of expense. In a very well built house which I lately had an opportunity of inspecting, the connection with the main was made with a ⅝-inch tap and *Another plan.*

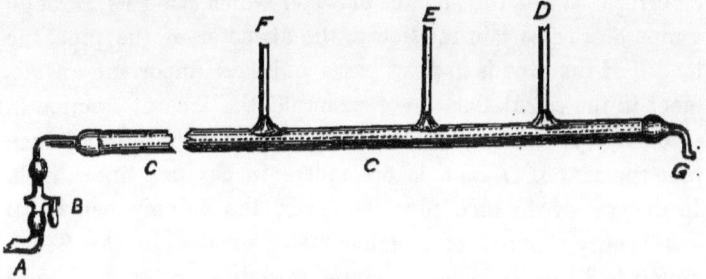

Fig. 18.

a ¾-inch **A A** pipe carried down below the frost line. This extended to the stop and waste cock just inside the cellar wall. From this point the water was carried to the rear wall of the cellar, some 50 or 60 feet, by means of a 2-inch **A A A.** From this a separate line of ⅝-inch **A A** pipe was carried up to each floor. In the drawing (Fig. 18), A is the connection with the main; B, the stop and waste cock; C, the 2-inch main service pipe; D, the supply for the kitchen and basement; E, the supply for the parlor floor, and F for the second floor. Though not in a very elevated position, a constant service could not be secured above the second floor, so the third and fourth floors, as well as all the water-closets in the house, were supplied from an elevated tank in a closet on the fourth floor, which was filled by a double-action pump with a vertical lever handle. Water was supplied to this pump by a 1-inch pipe connecting with the main service pipe at G. This system worked perfectly under all conditions, and has no objection except the considerable additional cost of pipe.

Its advantages explained. The theory of this arrangement is that by enlarging the pipe in the cellar we diminish the friction and facilitate the flow to such an extent that we secure a considerable increase in the amount of water available for house service. I frequently meet plumbers who find difficulty in understanding what advantage can be gained from using a large service pipe, as the amount of water which comes in through a five-eighths tap is, they insist, no greater than will flow through a five-eighths pipe. This is *Relation of length and diameter of pipe to discharge.* an error. While the amount of water which can pass through a pipe bears a certain relation to the diameter of the pipe, the length of the pipe is in most cases quite as important an element in the calculation. For example, the area of friction in a 1-inch pipe is 3·1 square inches per lineal inch. In a 2-inch pipe the area of friction is 6·2 square inches per lineal inch. In the case of the inch pipe, however, the volume per lineal inch is only about ·7 of a cubic inch; whereas in the 2-inch *Frictional resistance in small pipes.* pipe it is 3·1 cubic inches. Now, in passing a given volume of water through a pipe under a given pressure, it is evident that a much greater frictional resistance will be encountered in a 1-inch pipe than in a 2-inch, and that the aggregate of this *Short and long pipes.* resistance will be in proportion to the length of the pipe. It is a well-established principle in hydraulics that very much more water will flow through a short pipe than a long one under a

SERVICE PIPES AND WATER SERVICE IN CITY HOUSES. 119

given head. To state this exactly: With a given pressure the discharge is inversely as the square root of the length, or the length varies inversely as the square root of the quantity discharged. Thus, for discharges in the ratios of 1, 2 and 4 gallons, the lengths of pipe would be as 16, 4 and 1—that is to say, a pipe 16 feet long discharging 1 gallon in a given time, would, if 4 feet long, discharge 2 gallons in the same time, and if 1 foot long would discharge 4 gallons, the pressure in each case being the same. Applying this rule to the case under discussion, we may assume that by reducing the length of the five-eighths pipe connecting with the main to, say, 20 feet, we get a flow through it nearly twice as great as we should have were it carried back the whole 70 feet. To state the difference exactly, if 70 feet of five-eighths pipe discharged 4·47 gallons in a given time, 20 feet of five-eighths pipe would discharge 8·36 gallons in the same time, the pressure in both cases being the same. The shorter we can make our five-eighths pipe the more water we shall have, and if we could cut it down to one or two feet we should get as much water through it as 80 or 90 feet of 2½ to 3 inch pipe could carry away. Now, by increasing the diameter of the pipe we also increase its capacity in a very rapid ratio. With pipes of an equal length, having diameters of 1, 2 and 3, the quantity of discharge will be as 1, 5·6 and 15·6 respectively. In other words, a pipe to discharge three times as much as another pipe needs to be only a little more than one-half greater diameter. To be exact, a pipe 1·55 inch diameter will discharge three times as much water as a pipe of 1 inch diameter. It requires no further explanation to show that when we shorten our ⅝-inch pipe to the limit allowed by law and increase the diameter of the pipe leading from it, the flow of water will be enormously increased, notwithstanding the fact that we have only a five-eighths tap and connection with house to begin with. By making use of this principle, it is possible to pipe a house upon a plan very little more costly than that commonly followed in average city houses, and yet secure

Relative capacity of large and small pipes.

a much greater supply and a better distribution than is usually found.

Plates 1 and 2. In the illustration marked Plate 2, is shown the service-pipe system of a well-arranged city house, which will be better understood from an examination of Plate 1, showing the plan of each floor, with fixtures. The architect under whose direction the house was built has given special attention to the subjects of drainage and water service, and is one of the few in his profession who know how to distribute a water supply in such a way as to have a constant service at every cock. The drainage system of the house is shown in Plate 3.

Description of house. As shown in Plate 2, the house is of the usual style of city houses, consisting of a cellar, basement and three stories above ground level. There is also an extension in the rear two stories high and about half as wide as the main house. This extension contains a laundry with four tubs, servants' bath and water-closet on the basement floor and a dining room above. The fixtures in the house may be described by floors as follows (see Plate 1):

Fixtures *Basement.*—Two water-closets, four laundry tubs, one bath and one basin, besides the usual kitchen fixtures.

Parlor Floor.—A butler's pantry, with one sink and one basin.

Second Floor.—Two water-closets, one bath and five basins.

Third Floor.—One water-closet and four basins.

Main service pipes and branches. The water is brought to the house by a $\frac{5}{8}$-inch pipe, extending from the tap in the street main to the stop and waste cock inside the cellar wall. From this point it is enlarged to $1\frac{1}{4}$ inch, carried through the cellar and to the top of the kitchen boiler. All branches for a cold-water service at basins, baths, closets and wash tubs are made with $\frac{1}{2}$-inch pipe.

From the level of the boiler top the main service pipe is reduced to 1 inch, and so continues up to the level of the sec-
Reduction in size of pipe. ond story, all branches, as before stated, being one-half inch. At this point the main service pipe is reduced to three-quarter inch

and extends across the house under the second-story floor. As it rises to the third story it is reduced to five-eighths inch, and is then branched to supply two of the basins on that floor. The other basin is supplied by a ½-inch pipe, carried up from the 1-inch pipe on the floor below.

The distribution of the hot water is effected in the same way. Following the general line of the cold-water pipe, as described, it begins at the boiler with 1 inch diameter and is carried to the level of the second story. As it turns to cross the second-story floor it is reduced to three-quarter inch, all branches being one-half inch, as in the case of the cold-water system. All the fixtures below the level of the boiler are supplied through a ½-inch line, which is branched in the usual way. This gives the smallest amount of main hot water service pipe on the floor where the pressure is greatest, while on the upper floors the diameter is greater to offset the diminished pressure. *Hot water distribution.*

The circulating pipe extends from the 1-inch pipe on the second floor down to the branch which ends at the sediment cock. This pipe is of one-half inch diameter throughout. *Circulation.*

It will be seen at a glance that the size of the main service pipes is in proportion to the quantity of water to be delivered beyond any given point. The relative capacity of the different sizes of pipe used may be tabulated as follows, the figures opposite the diameters showing their area in inches and decimals:

Diameter, inches.	Area, inches.	Diameter, inches.	Area, inches.
1¼	1·22	⅝	·30
1	·78	½	·19
¾	·44		

Relative capacities of pipe.

In going downward from a given point the reduction would necessarily be very rapid, because the constant increase in the pressure would have to be equalized by an increased friction. This explains why, in the basement, a half-inch line is all that is needed to supply hot water to all the fixtures.

The practical plumber does not need to be told that it is but little, if any, more expensive to pipe a house in this way than

122 SERVICE PIPES AND WATER SERVICE IN CITY HOUSES.

Advantages. by the method commonly employed. The advantages, however, consist in the difference between a distribution which will permit every cock in the house to run full bore at the same time, and one in which a flow at one point cuts off the supply from points in the main line and its branches above or beyond. The plan we have described admits of such modification as will adapt it as well to one style of house as to another. Its essential features are:

Essential features of the plan. 1. Where the size of tap and connection with house is prescribed by law, making said connection as short as possible.

2. Increasing the diameter of the pipe as soon as the law permits, and carrying it without reduction beyond the point where the greatest quantity of water is needed.

3. For the sake of economy, reducing the pipe gradually as the number of branches to be supplied diminishes.

4. Making branches as small as will deliver the quantity of water needed. In most cases one-half inch is sufficient, and in some cases three-eighths will answer.

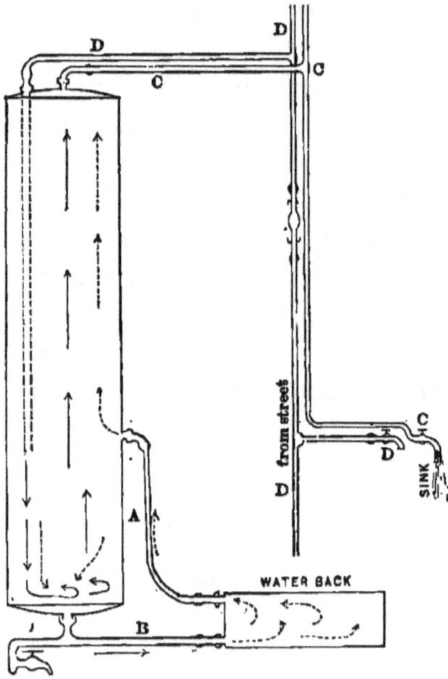

The maintenance of an abundant service of hot water through a house presents few difficulties to the practical plumber, and the suggestions which I shall have to offer on this point will occupy but little space. The theory of heating water in a boiler by passing it through the water back of a range is generally well understood, but may be briefly described for the

Hot-water service.

The theory of the water back.

Fig. 19.

SERVICE PIPES AND WATER SERVICE IN CITY HOUSES. 123

information of non-professional readers. In the illustration marked Fig. 19, the boiler is fed with cold water through the pipe D, entering at the top and extending down inside the boiler as indicated by the dotted lines. Consequently, all the water which comes into the boiler cold is discharged into it near the bottom. Being heavier than warm water, it remains at the bottom. Through the pipe B the water is supplied to the water back of the range, and thence passes back into the boiler through the pipe A. The outflow from the boiler to the hot-water cocks is through the pipe C, connecting with the top of the boiler. Without applying heat to the water back, we should have merely a boiler with its connections, including the water back, full of cold water, which, of course, would stand motionless. When we apply heat to the water back, however, we compel a circulation throughout the boiler, the water back and the pipes connecting them. The reason for this enforced circulation will be seen at a glance. When water is heated it has the same tendency to rise which air manifests under like conditions. In a closed vessel we should only have ebullition, more or less violent according to the degree of heat imparted to the water. In the case of a water back, however, which has an inlet and an outlet, both below the line of the head, the heated water in its upward flow passes out through the pipe connecting with the boiler and is replaced by water from the boiler—its course in the water back being indicated in the drawing by the curved arrows. Thus a constant circulation is maintained into, through and out of the heated water back, which is slow or rapid in proportion to the rapidity with which a high temperature is imparted to the water. *Circulation.*

It will be seen from an examination of Fig. 19 that the pipe which carries the water from the boiler to the water back is connected with the bottom of the boiler, while that which carries the water from the water back to the boiler discharges into the upper part of the boiler. This is done to insure the maintenance of a circulation in the boiler. By placing the hand upon a boiler when a fire in a range or stove maintains a circu- *Boiler connections.*

lation through the water back, it will be seen that the upper part is hot and the lower part cool—or at least much cooler than the upper part. As the outflow from the boiler to the various cocks connected with the hot-water service pipe is always from the top, it is obvious that the supply is drawn from the warmest stratum of water in the boiler. Now if, as we find in private houses scientifically piped, the pipe which carries the hot water from the top of the boiler through the house is brought back and again connected with the boiler at or near the bottom, a circulation of hot water is maintained throughout the house. The hot water rising within the boiler is replaced by cold or cool water flowing into it at the bottom, and is carried into and along the pipe by the pressure of hot water rising behind it. The pipe by which this circulation throughout the house is maintained is commonly of small size. Under average conditions half-inch is large enough. The way in which this pipe is connected with the hot service pipe and the boiler is shown in Plate 2.

Circulating pipes.

Size of circulating pipes.

Advantages of circulating pipes.
The advantages of a circulating pipe are twofold. It permits the water cooled by the radiation of its heat as it passes through the hot service pipe to return to the boiler, thus making room for the hot water which follows it. This keeps the pipe from becoming cold, and obviates the necessity for emptying it when warm water is wanted on one of the upper floors. It also maintains a more equable temperature in the boiler, and tends to relieve it from undue pressure. In proportion as the temperature of the water in the boiler is raised, its flow through the circulating pipe is rapid, and the loss of its heat by radiation from the pipe will usually keep the temperature in the boiler below 212° Fahr.

Sags.
In providing for such a circulation, care must be taken to guard against sags. If the pipe drops out of line its own diameter, the circulation will be very slow and uncertain, if it is not stopped altogether. It is, therefore, especially desirable to give hot-water pipes, if lead or tin lined lead, continuous sup-

SERVICE PIPES AND WATER SERVICE IN CITY HOUSES. 125

port when carried under floors. If laid upon shelves, any stretch due to expansion will be provided for without interfering with the circulation.

In connecting a boiler with a water back, it is a good plan to make the upper pipe a little larger than the lower. For example, if ⅝-inch pipe is employed to carry water from the boiler to the water back, ¾-inch pipe should be used to carry the water from the water back to the boiler. This gives a free flow and promotes an active circulation. It is customary, I believe, to make both pipes the same size, but an important advantage will be found to attend the use of two sizes of pipe, as suggested. *Connecting boiler with water back.*

The boilers most generally employed in good work are of copper, tinned on the inside. These are probably the best, as they are lighter, and stronger in proportion to weight, than those made of any other available metal. Black iron boilers, though used to a limited extent in some cities, are no longer employed in the best plumbing practice, and galvanized iron, though somewhat extensively used, are not to be recommended for reasons elsewhere given in detail. Kitchen boilers are not usually subject to dangers of any kind, but prudence suggests the adoption of precautions against explosion and collapse. When supplied from tanks, an air pipe carried up to and bending over the tank gives the requisite security. When filled direct from the mains this precaution cannot be taken—unless it is possible to carry the pipe above the point to which the water would rise under any possible condition of head or pressure. For such cases vacuum and safety valves are made, but they are not calculated to inspire the impartial mind with unlimited confidence in their efficiency. They are made of brass, and as they are commonly allowed to remain untouched for years, they are very liable to corrode fast in their seats and, in my experience, are seldom ready for prompt action at the moment of emergency. Probably they are of little practical use. I never heard of but one authenticated case of explosion in a kitchen boiler in this country, and that took place under *Boilers. Accidents to boilers. Vacuum and safety valves. Explosions.*

126 SERVICE PIPES AND WATER SERVICE IN CITY HOUSES.

Collapses. very peculiar circumstances. Collapses are more frequent, but are rarely attended with any worse consequences than those produced upon the boilers. They result from the creation of a vacuum in the boiler by the condensation of steam when there is a sudden and strong inflow of cold water. If, when water is shut off from the house from any cause, the hot-water cock nearest the boiler is opened, and left open until the supply is turned on again, there is no danger of collapse. I have no desire to discourage the use of safety and vacuum valves, but if used and depended upon, they need to be frequently looked after to see that they are in good working order.

Water from boilers. Generally speaking, water drawn from kitchen boilers should not be used in culinary operations. Indeed, there would be very little temptation to do so if the condition of the inside of boilers could be seen. They are commonly deeply incrusted *Water cocks.* and usually very foul. In good work a waste pipe with a cock is provided, to empty the boiler and discharge the sediment. I have no doubt that a great deal of foulness, which would otherwise be retained, passes out when the sediment cock is opened, but from the condition of boilers I have seen taken down and opened, I am very certain that vastly more remains in than passes out. If the pipe which supplies cold water was closed at the end and perforated for six or eight inches, it would fill the boiler equally well and wash it out a great deal better. This arrangement is shown in Fig. 20.

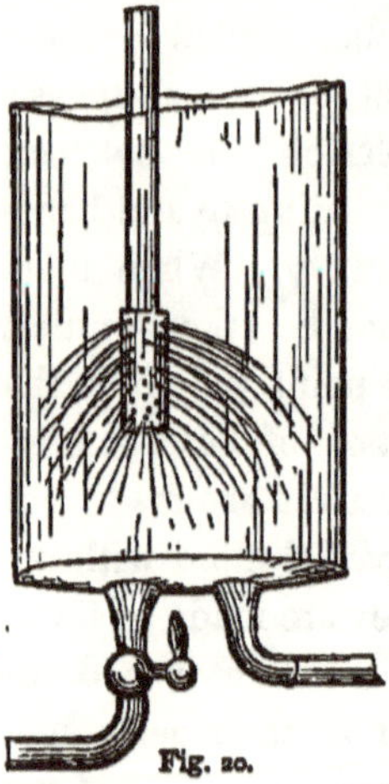

Device for cleansing boilers.

Cocks.

Fig. 20.

Of cocks, or faucets, it is scarcely necessary to speak particularly. They are made in almost unlimited variety, and there are a great number of patent cocks, some of them self-closing, which are competing on about equal terms for popular favor. *Relation of cost and quality.* Concerning them all it is only necessary to say that the best are, as the rule, the most economical. The effort of a majority

SERVICE PIPES AND WATER SERVICE IN CITY HOUSES.   127

of makers now seems to be to compete in cheapness, rather
than in excellence, and as the consequence the market is flooded
with cheap cocks in uncounted variety, which are very flimsy
affairs. Those who want a good article, however, can always
find it if they buy of responsible dealers and pay the price of
good materials and workmanship.

In piping a house, it is of the utmost importance that intelli- *Protecting
gent measures should be taken to protect the service pipes from *service pipes from frost.*
frost. Nearly one-half of our country is every winter exposed
to temperatures ranging from zero down. Searching winds
aggravate the discomfort of the cold and penetrate every
corner of our houses. For several months of every year we *Houses not built for cold weather.*
have in the Northern States a semi-arctic climate, and as we do
not commonly build our houses with a view to making them
comfortably habitable in very cold weather, frequent stoppages
of water service from the ice in the pipes must be expected.
The houses where the plumber is not needed after every *Winter work for plumbers.*
"cold snap" are few and far between. Plumbing is often
done as if summer was expected to last through the whole year.
For this reason the plumber receives in cold weather a larger
share of honest, hearty abuse than any other mechanic connected with the building trades. I am not prepared to say that
much of this abuse is not well merited, for a great deal of the
trouble experienced in winter from the freezing and bursting *Why pipes burst.*
of service pipes is due to bad workmanship or to ignorance—
or both. In New York the trouble begins about Christmas,
and unless the season is unusually mild, it grows steadily worse
until the warm spring sun brings relief, when the long interrupted flow is resumed and the plumber comes to do his final
patching up. During and immediately after a "cold snap" in
January or February, our streets present a curious appearance.
At frequent intervals we see groups of laborers tearing up the *Experiences in New York.*
pavement, or building bonfires to thaw the ground enough to
permit them to expose the buried pipes. During the winter of
1874 I saw as many as a dozen such groups on one block in the

upper part of New York, and while the effect upon the eye was picturesque, the effect upon those compelled to pay over and over again for the same work, which never should have been necessary in the first instance, was not all that could be desired. Probably we shall see the same thing every cold winter for years to come, and while the public misfortune will bring profit to the plumbers, it will not benefit the trade in the long run.

Sanitary evils of an interrupted water service. Regarded from a sanitary standpoint, the evils of depriving a city house of its water service are almost incalculable. The first and most important of these is the impossibility of flushing the water-closets and renewing the seals in waste-pipe traps. *Foul water-closets.* The inconvenience suffered, great as it may be, is of small consequence compared to the danger of sewer-gas poisoning (especially in the absence of adequate soil-pipe ventilation), and the nuisance resulting from the accumulation of foul *Secondary decomposition.* matter in water-closets. A mass of healthy human excrement is an offensive, but not necessarily a dangerous, nuisance at first, but it commonly becomes so in from 36 to 48 hours, and when houses are without water for a week or two at a time, the unclean water-closets may become the source of widespread infection. When water has to be brought into a house by the bucketful, very little of it is commonly used for flushing water-closets or for the replacement of seals in waste-pipe traps, which, in the absence of a regular water supply, have *Influences which affect the security of water seals in winter.* been rendered foul by the absorption of sewer gases. Winter is the season in which the air in sewers acts with greatest pressure upon the seals in traps, owing to the closing of the ventilating holes in manhole covers by ice, snow and mud. It is at this season that our traps and seals are least able to withstand the effects of this back pressure from the sewers.

Service pipes freeze from causes easily recognized and almost always remediable. Commonly, cheap workmanship and ignorant planning are to blame. Plumbers may do a great deal toward preventing this freezing; first, by giving their patrons

SERVICE PIPES AND WATER SERVICE IN CITY HOUSES. 129

directions as to the best methods of protecting individual houses according to the circumstances of the case; and, second, if they have a job to put up, doing it in such a way that freezing of the water in service pipes is guarded against. There are a great many houses where the plumber, by a little careful planning and the putting in of a few safeguards in the way of waste cocks, or by jacketing pipes with a non-conducting coat, can prevent the water supply from being cut off by frost. <small>Jacketing pipes.</small>

To prevent the freezing of water in exposed pipes, recourse is commonly had to the expedient of maintaining a constant flow from indoor cocks. A circulation is thus kept up which, in any but extremely cold weather, will keep the pipes open. This is a very easy expedient for the individual householder, but when it is followed in half the houses of a large city during the greater part of a cold winter, the expense to the public treasury involved in paying for the enormous resulting water waste is a very serious matter. Few people who have not calculated it have any idea of the amount of water which will run to waste in one night in a stream no larger than is usually considered necessary to keep a pipe open in frosty weather. A stream no larger in diameter than a lead pencil and running silently, will easily flow at the rate of 700 gallons in 24 hours. With a head equal to about 30 feet, I have seen an ordinary basin cock deliver water at the rate of a gallon in three-quarters of a minute. This would amount to considerably more than 1400 gallons per day—a quantity more than sufficient for all the wants of a family of 20 persons. This waste costs the taxpayers a great deal of money, and will probably continue wherever water meters are not used. Granting that such a waste of water will commonly keep pipes open, it must be confessed that there is no excuse for a condition of affairs which renders it necessary. An architect and plumber who cannot together pipe a city house in such a way that no interruption of the water service from frost need be apprehended until the <small>Water waste in cold weather.</small> <small>The flow of small streams.</small>

130 SERVICE PIPES AND WATER SERVICE IN CITY HOUSES.

street mains shall freeze, cannot be considered masters of their respective professions.

Freezing of pipes under ground. One of the points at which freezing occasions much trouble and expense is between the main and the foundation of the house. This section of the service pipe can, as the rule, be reached only by digging up the street or tunneling from the cellar out, both of which are troublesome and expensive. In such cases it is sometimes possible to steam out the pipe by means of an apparatus made for the purpose, but more or less excavation is generally necessary, with the inevitable accom-

Fig. 21.

paniment of a "nasty muss" in front of one's door. Street mains, especially in rocky soils, are often laid within 3 feet of the surface. When the service pipe is carried straight to the house from a main thus situated, it is pretty sure to freeze solid in very cold weather. Weeks will pass before it thaws out in the natural way, and if the plumber attempts it he is likely to find it a costly and troublesome job. The best way to protect a pipe at this point, and the one which is employed by some of our best plumbers, is shown in Fig. 21. The main, which is supposed to be 12-inch, is within about 3 feet of the surface,

Difficulty of thawing buried pipes.

or about 2½ feet from the paving stones. After connection is made with the main, the pipe is carried down 24 inches, or about 5 feet from the surface of the street. The vertical part of the pipe and the tap are covered by a wooden box, which comes about 5 inches above the top of the main. The box is made about 18 inches square, and filled with concrete, mortar, tan-bark, or any substance which will exclude the frost. The horizontal part of the pipe is protected by its depth below the surface. When this simple precaution is taken there will be no occasion for digging up the street until the main shall freeze. *Boxing.*

It is a good plan to carry the service pipe in under the cellar floor, with a large stop cock near the foundation wall. When branches rise against the side walls they should be provided with waste cocks, and should be boxed and packed as far up as may be needed to protect them from the frost. *Carrying service into houses.*

When for any reason it is difficult to go deep enough to carry the buried pipe below the frost line, it is a good plan to inclose it in an iron pipe large enough to leave an air space all around. In this case it is very easily and quickly thawed out with steam. *Encasing buried pipes in larger ones.*

In plumbing warehouses, factories, and other large buildings with deep cellars and sub-cellars, it is often necessary to carry the service pipe across open areas protected only by gratings, which are left uncovered to admit light. When this is necessary, the pipes should be warmly jacketed in some way—either by boxing in sawdust, or wrapping with felt or other thick, porous fabric. Inside a house they should be carried up in some warm, sheltered corner, where they will never be exposed to a freezing temperature. As already stated, stop and waste cocks should be provided, and the pipes should be so laid and supported that they can be drained dry. In buildings improperly planned, it is not always possible for the plumber to correct the original mistakes of the architect without undertaking more extensive and costly alterations than the owner or tenant *Jacketing pipes which traverse open spaces.*

is commonly willing to pay for. He can, however, do much in this direction by adopting the precautions suggested. There are many ways in which freezing and bursting can be guarded against. The first and best is to have the pipes taken from cold corners and put in warmer positions, where they will not freeze. This will cost something, but it is cheaper in the end than paying for frequent repairs. This remedy is positive and effectual. When this cannot be done, a little contriving will enable the plumber to stop up holes, cover the pipes and thus protect them in all but the most exceptionally cold weather, when recourse must be had to the expedient of emptying them. Sometimes they freeze when apparently well placed, owing to openings between joints or cracks in woodwork, which convert the pipe boxes or the niches in which they are placed into cold ventilating shafts. Such openings must be carefully closed. When a current of cold air follows the pipes up from the cellar, they will probably freeze however placed.

I have seen pipes in exposed positions effectually protected from frost during long and very cold winters, by bedding them in plaster of Paris, and have several times tried this plan where others had failed, with entire success. The material is inexpensive, it is easily applied and presents no difficulties or objections.

Of the treatment of frozen pipes I shall speak but briefly. Every plumber is familiar with the expedient of sending steam through a small rubber tube introduced into a frozen pipe. To do this successfully a good boiler is needed, with safety valve, and a fire lively enough to give a good supply of steam. Some plumbers consider hot water better than steam for this kind of work, being more rapid and convenient. Nothing is needed but a kettle over the fire, the usual rubber tube and a pump. The pump should have brass valves, and the piston or sucker must be packed with cotton, as the heat of the water will destroy leather valves and packing. With such a pump a stream of hot water can be sent directly in upon the ice. While an equal

SERVICE PIPES AND WATER SERVICE IN CITY HOUSES. 133

weight of live steam is many times more effective than hot water, the amount of steam that can be obtained from one of the boilers used by plumbers is so small, and the loss from condensation before the steam reaches the ice so great, that a steady stream of hot water does the work much more rapidly. The weight of hot water sent through the pipe can be made, per- <small>Hot water.</small> haps, twenty times greater than that of the steam which the same boiler and fire would furnish. In most cases, too, the water collects in the steam pipe, and the ice is, after all, only thawed by a small stream of hot water heated and driven by the steam. The difficulty which bends and angles make in the use <small>Bends and</small> of this system prevents its adoption in a considerable number <small>angles.</small> of cases.

When excavation is necessary, the quickest and cheapest <small>Excavation.</small> mode of procedure in dealing with a pipe leading to the street main is to employ laborers accustomed to the work, and dig down to the pipe in two places. Then with small shovels tunnel <small>Tunnelling.</small> each way, cutting the earth away under the pipe. Then heat up with shavings, straw or other light material. When the <small>Thawing.</small> pipe is thawed out, it will sink to the bottom of the tunnel, and by this means will probably dip below the frost line. The tunnel is then filled with earth, rammed in to insure solidity, and the pipe will probably be safe for the future.

These methods of dealing with frozen service pipes are appli- <small>Thawing</small> cable only to lead, tin and tin-lined pipes. With iron pipes, pump <small>iron pipes.</small> barrels, steam coils and the like, the external application of hot water, steam or the heat of a fire would be very dangerous. In thawing such pipes it is best to employ cold water. The opera- <small>Cold water.</small> tion is usually tedious, but it averts the danger of a burst, which is of more consequence in the case of an iron pipe than in that of a soft metal which can easily be repaired by making a solder joint.

When it is impossible to prevent the freezing of pipes, we <small>Guarding against bursting of frozen pipes.</small> have one other alternative, and that is to fix them so that they shall not burst if they freeze. There are two or three ways of doing this. The expansion of water in freezing tends to force it along the pipe, and does not exert any considerable pressure upon the sides until the whole body of water is caught in such a way that it must freeze where it is; then the expansion begins to make itself felt. It is on this account that pipes in which the cocks are closed by a spring against the water pressure, can so frequently be frozen without bursting. As the pressure increases the water finds an outlet. Wishing on one occasion to prevent the bursting of frozen pipes, I took a short length of large iron pipe, closed one end with a cap, and placed inside a three inch rubber ball. This little air chamber was then *Rubber balls.* attached to the service pipe by a collar. The collapse of the ball under pressure made room for the expanding water, and so protected the pipes from injury. This precaution is ineffectual at times, because the pipes may be "caught" by the frost at two different places at the same time and the water confined between them so as to burst the pipe. If, however, a small rubber tube be carried all through the service pipe, freezing would be rendered harmless. The smallest size of rubber tubing made, which has about one-eighth inch bore, would answer perfectly, and if the ends were taken outside the service *Continuous* pipe this might be made a means of thawing out the pipe. This *air chambers.* pipe is closed at the ends when they are not taken out, and forms a continuous air chamber, the water flowing around it. This plan has been proposed a great many times, but I do not know that it has ever been put in practice on any considerable scale. An ordinary air chamber would answer the purpose very well, except for the fact that the air would soon be absorbed by the water, and then the chamber would be useless.

As the bursting of pipes by ice is a phenomenon not generally understood by plumbers, a few words on the subject may be found of interest. When a substance solidifies or freezes, there is always a change of volume, which usually is a contraction; but in the case of water and a few metals, such as cast iron, antimony, bismuth, &c., an expansion takes place. The expansion of water at the freezing point is by no means gradual, but it takes place almost instantaneously, and the force exerted at the time is enormous. Of the amount of force exerted by freezing water we have ample testimony in the bursting of our water pipes and other sadder calamities; but it is a very simple matter to calculate pretty accurately the amount of force developed. Grassi proved the compression of water to be proportional to the pressure; he also found from very accurate experiments that with a force of one atmosphere, or 15 lbs., water was compressed 0·0000503 of its original volume; and knowing that water exerts in expanding a certain volume a force equal to the one required to condense it to that amount, the following simple calculation will show pretty accurately the amount of force exerted by water when freezing:

$$0\cdot0000503 : 0\cdot1 :: 15 \text{ lbs.} : 29{,}821 \text{ lbs.}$$

That is to say, in freezing, water exerts a pressure of about 30,000 lbs. per square inch, which far surpasses the strength of our stoutest pipes; and if this enormous power is exerted on the pipe instead of being expended in compressing something which can yield without fracture, it is not to be wondered at that the pipe bursts.

The quality of water drawn from the best available sources of supply is often such as to require filtration. On this point a few suggestions may be of use to the general reader. The man who buys a filter commonly labors under the impression that the water which passes through it will leave all its impurities behind. He uses it for a few weeks and probably finds his expectations realized, so far as he can judge from appearances. It works so well that he gives it no attention. Presently he finds that the water is as bad as before he got his filter, and perhaps worse. Then the filter is condemned and taken out, and no one need talk filters to him again. The truth is that he started out with a mistaken notion, and did not give his filter a fair chance to do its work. It cannot be left to retain the impurities it removes from the water passing through it without itself becoming foul.

<small>Filtration.</small>

<small>The popular idea of a filter</small>

136   SERVICE PIPES AND WATER SERVICE IN CITY HOUSES.

Impurities in water. The impurities in water may be classed under two heads, the mineral, or fixed impurities, and the organic, which may be animal or vegetable, or both. Under the first head may be classed all those substances which do not undergo decomposition, as sand, clay, chalk, &c. The operation of these substances is simply to stop the filter's action by making it impervious to water, the quality of the water changing very little even when the filter is well filled with them. With vegetable and animal matter the case is different. Arrested at the filter, these substances undergo fermentation, more or less rapid according to the temperature and other circumstances. The products of the decomposition of many of them are soluble in the water and, consequently, as they are dissolved pass easily through the filter and make the water even worse for drinking than it was before filtration. Take, for example, Croton water during the summer months and in the early spring. In the lower wards of New York it is found to contain sand, infusorial shells, a small amount of vegetable matter, and a small amount of animal matter, part of which is easily decomposed. *Foul filters.* The result is that when any amount of this matter accumulates in the filter it begins to decompose, and is quickly carried through the filter either in a dissolved state or else in a more finely divided state. At first the filter acts well, but the more impurities it arrests the worse the result, for as decomposition sets in there is a greater quantity of matter to be acted upon. This decomposition of substances arrested by the filter is always a source of more or less danger—easily avoided, however, but one which must be provided for if pure water is to be delivered continuously for any length of time. It will be understood that these remarks apply more especially to those filters which deliver but small quantities of water, and not to the large ones such as are used for water works.

Filters like sieves. If it is desired to make a filter practically efficient, we must pursue the same plan we should follow in screening gravel through a sieve. After a certain amount has accumulated which will not pass through the meshes, the operation of screening is suspended and the sieve emptied. If this were neglected too long the sieve would become choked and useless. For the same reason, a filter must be so constructed that it can be washed clean, or the filtering medium removed and replaced with new. When sand is the medium, this is easily and quickly done. Sponges and like materials can be washed, and if this is thoroughly done they are as good as new and can be used a long time. Charcoal is not easily purified, but is easily replaced with fresh. It is valuable because of its vast powers of absorbing and destroying organic matter. Artificial stones, sandstones, and other porous materials of a similar nature, are good for a little while, but if under pressure are very liable to crack and become useless. The fact that they are costly and cannot be easily and thoroughly cleaned, is a great drawback to their use. Cotton cloth, closely packed, especially if old and fine, would probably answer a very good purpose, and would have the advantage of being easily and quickly washed. For convenience and efficiency, probably there is no material equal to sponge, tightly packed. It presents a vast amount of surface in a very small compass, and is very elastic and durable.

    Another point to be noted is that it is very difficult to filter water when it is forced through the apparatus under heavy pressure and at great speed. It should come through slowly. A foot or two of head is enough, and the quantity must not be large, as the pores of the material would have to be so large as to allow passage for dirt as well. A very fair sort of a filter can be made from two 6-inch flower pots. Fill one with sand and gravel, stopping the hole loosely, so as to obstruct the passage of the sand. Pack the other with sponge washed clean. It will take considerable, but if much cannot be afforded use a smaller pot, and put it in tightly to prevent the water flowing through between the pieces. Set the pot filled with sand and gravel on top of the other, and allow the water to run through both. Such a filter will give a good amount of clear water and will not require much washing. This will depend, of course, entirely upon the character of the water. If it is only clay or mud, the accumulation of it will simply diminish the flow, while organic matters must, as I have shown, be removed very frequently. Even a cupful of sponge, tightly packed, will strain Croton beautifully, and probably remove the worst of the impurities, but it will require to be washed often. A little ingenuity will enable anyone to devise a convenient and efficient filter, and if too much is not expected of it, one which will accomplish the object sought. The market is well supplied with them, and most of those offered for sale will do good service for a time if properly managed.

Sponge strainers

CHAPTER VII.

TANKS AND CISTERNS.

In city houses it is sometimes necessary to raise the water needed to maintain a constant service at water-closets, cocks, &c., into tanks or cisterns. The reason for this is sufficiently set forth in the preceding chapter; the methods employed in raising water will be discussed in Chapter IX. Here we shall consider briefly the vessels or reservoirs provided for the storage of water in city and country houses, and their proper care. *Tanks in city houses.*

In constructing a tank for the storage of water in a city house, it is necessary to observe several conditions, none of which can safely be disregarded. For convenience it should be of size sufficient to supply the average consumption during at least twelve hours. This is not imperative, perhaps, for it can be filled as often as may be necessary; but if the work of pumping is done by hand, it is likely to be neglected except at fixed periods. The tank should, therefore, be so large that it is not likely to be emptied before night if filled in the morning. Comfort and decency, if not health, demand that water-closets should never remain foul because there is no water to flush them; also, that personal ablutions and the toilet should not be interrupted nor interfered with by the want of water. *Construction. Size and capacity. Tanks should never be empty.*

When the water stored is to be used only for washing, flushing water-closets, and for any and all other purposes except drinking and cooking, it makes practically but little difference what the tank is made of, or lined with, provided it be strong, tight and durable. In houses occupied by single families, water from tanks is rarely used in cooking or on the table. A service is almost always maintained from the street mains at the cocks on the lower floors, and there is no need of using the water which has stood in the tank for culinary or table pur- *Tank water not usually needed for cookery or drinking.*

140   TANKS AND CISTERNS.

Exceptions. poses. In tenements, flats, warehouses, office buildings, &c., the water drawn on the upper floors is generally taken from the tanks. In such buildings it is necessary, and in all buildings it is desirable, that tanks should be made of some material which will not render the water held in them unfit to be taken into the stomach.

Wooden tanks lined with tinned copper. The best tanks which I have seen for use in city houses are strong wooden boxes lined with tinned copper. Of this material there are two kinds, and between them there is an important difference. *Sal ammoniac and resin as fluxes.* In one kind *sal ammoniac* is used as the flux in tinning; in the other resin is used for this purpose. My observation leads me to the opinion that the latter only should be used for tank linings, for the reason that tin put on with *sal ammoniac* does not retain its hold upon a copper surface under water, and a corrosive action is likely to take place which will rapidly destroy both the tin and the copper. *Superiority of tinning put on with resin.* The resin tinned plates cost a little more than those tinned with *sal ammoniac*, but in durability and security against metallic poisoning, there is no comparison between them.

Wrought-iron tanks. A very good form of water tank is made of rolled iron plates, riveted up and lined with cement. If the proper quality of *Cement linings.* cement is used—such, for example, as is placed in the bottoms of iron ships or on the bottoms of large aquaria—and carefully applied, it constitutes an insoluble and impervious coating which will last for many years. *Purification of water by contact with iron.* Iron tanks are commonly used without such lining; and while the water is, if anything, purified and rendered more wholesome by direct contact with the iron, *Rust.* the presence of too much iron rust renders it undesirable for household purposes. This consideration applies especially to small tanks containing a supply only equal to the daily consumption of one or two families. *Discoloration of water by rust greatest in small tanks.* When the tanks are of great size, as in hotels and other large buildings, the surface of iron exposed is so much less, in proportion to the cubic contents of the tank, than in small ones, that no perceptible discoloration from rust is likely to occur. *Objection to small iron tanks.* The objection to small iron tanks is chiefly found in the fact that the level of the water in them is constantly fluctuating, and that they are drawn upon as freely when but little remains and that little charged with rust from the bottom and sides, as when full. Any impervious and insoluble coating of non-poisonous composition which will protect the metal from rust, will render an iron tank all that can be desired. Linings for iron tanks.

The use of lead as a lining for water tanks cannot be recommended under any but exceptional conditions. What these conditions are need not be considered here, as the action upon lead of potable waters of various compositions is discussed at length in the chapter following. Untinned copper, yellow metal, galvanized iron and sheet zinc should not be used. Block tin is good, except under peculiar conditions, but owing to its cost it is not likely to be extensively employed. Other things being equal, a metal which can be used without having exposed soldered seams, is better than one which cannot. If kept full and clean, tanks of cedar, cucumber wood, white pine and certain other woods, without metallic linings, will be found to answer the purpose very well. Lead tanks. Action of water on lead. Copper, brass and zinc. Block tin. Soldered seams. Wooden tanks without linings.

Tanks should be kept clean. This is so evident a truism that the reader may smile at seeing it stated thus seriously; but I have found, from somewhat extensive observation, that the duty of cleaning them is commonly neglected—often for long periods. As the rule they are placed in inaccessible positions, and are rarely examined for any other purpose than to see how much water they contain. The dust which falls into them, as well as many of the impurities of the water, settles to the bottom and clings to the sides, and as tanks are usually so constructed that they cannot completely empty themselves through the service pipe, they sometimes accumulate a great deal of dirt of very miscellaneous composition. This almost invariably occurs in houses divided into "French flats," or let by floors to several families. In houses of the former class, the tank is usually under the care of a janitor, who probably pays no further attention to it than to see that it is full; in houses let by floors Tanks should be kept clean. Neglect. Sediment. Tanks in tenements and apartment houses.

or apartments, the condition of the tank is the business of no one in particular, and it is seldom looked at. Not long ago I was prompted by curiosity to examine the tank from which was drawn all the water used by two families occupying the two upper flats in a very elegant building in New York. It was filled and drained through lead pipes, which is not to be wondered at, as the employment of lead is usual in plumbing work. The water was raised by a pump in the cellar, operated by a small caloric engine. The tank was a wooden box lined with sheet zinc soldered at the corners, and between what may be called the high and low water levels—the tank was never quite full nor quite empty—the metal was much corroded. The color of the bottom and sides was so peculiar that I climbed into it, with great difficulty, after drawing off the water, and found that the bottom was covered with a layer of soft, oozy mud two inches deep. A similar coating covered the sides to a hight of twelve or fifteen inches. This mud, of which nearly half a barrel was removed, I found, upon examination under the microscope, to be composed largely of earthy sediment and organic matter. From the fibrous appearance of the latter, I concluded that it was principally the lint of carpets, drawn up by the ascending current through the well-hole into which the room containing the tank was ventilated. The thick deposit of fluffy dust on the edge of the tank indicated the amount which must have fallen into it. Exposed to the air for twenty-four hours, this composite mud became offensive to the sense of smell, and by the end of forty-eight hours, when partially dried by evaporation, it had become mouldy. In the eighteen months during which the house had been occupied, no one had thought of examining the tank, and the two families had continually drank the water decanted from this unwholesome sediment without even filtering it. In all houses supplied from tanks there should be pumps upon each floor occupied by a family, to lift, from the level to which its head will carry it, all the water needed for drinking or cooking.

But the sediment was not the only evil discovered during my inspection of the tank I have just mentioned. The floor on which it was placed was divided into a drying room, three bedrooms for servants, and the tank room, in which was also the servants' water-closet. They stood side by side, and the water-closet was as bad as one could be. It was a cheap, loose-jointed pan closet, so much out of order that half enough water to seal it would cause the pan to drop. The soil pipe of the house ended at this closet, and was without ventilation. As a consequence, the closet was extremely offensive, and the door of the tank room was kept closed by the servants so that the smell should not penetrate to their bedrooms. From what has already been said in the preceding chapters of the capacity of water for absorbing gaseous impurities, it is evident that this startling juxtaposition of foul water-closet and water tank was an evidence of the utter disregard of sanitary conditions which characterizes much of the contract building of the present time. *Conjunction of water-closet and tank.* *Dangerous conditions.*

In the construction of all kinds of cisterns and tanks for the storage of water used for drinking (and it may be assumed that the water constituting the principal supply of a house will sometimes be drank, or used by careless servants in the kitchen, however impure it may be), it is of the utmost importance to guard against the poisoning of the water by the gases of decomposing organic matter in sewers and cesspools. It is a well understood and generally accepted fact that the poisoning of water by sewage or the gaseous emanations therefrom, is a fruitful source of sickness. The poisoning of the water in tanks may occur in a variety of ways. Gases of a poisonous character may find their way into tanks through overflow pipes carried into soil pipes and, under some circumstances, through service pipes. These dangers are easily avoided by an intelligent arrangement of the plumbing work. The only way to keep dust out of tanks seems to be to cover them. Covers, if used, should be light, tight-jointed and easily removed, and ventilation should be secured by means of air pipes. Water, when *Water in tanks should be free from contamination.* *Organic impurities from sewer gas.* *Covers.* *Ventilation.*

144   TANKS AND CISTERNS.

Purification of water in tanks. stored in casks or other closed vessels, seems to undergo a sort of purifying process, by which many of its impurities are thrown down as a sediment. Sailors assert that water clears itself by "working," after the manner of wine or liquors. This is not strictly true, but it is a fact that most of the impurities held mechanically suspended in water are thrown down under these conditions, and it becomes really purer and more wholesome than when fresh. If we can exclude dust and give sufficient time for settling, water is pretty sure to be improved in quality, unless contaminated by contact with the vessel containing it.

Cisterns for rain water. In this country, cisterns in the rural districts are commonly built underground, of brick and cement. Many country houses *Objections to underground cisterns.* have these large underground cisterns, and are supposed to be provided with every convenience necessary in the way of water supply. Yet instead of having to pump water into the house and carry it about in pails, it should run in and distribute itself in pipes. There is no difficulty in securing this. *Elevated cisterns.* Small tanks, supported at a sufficient elevation to give the required head, are as easily filled from roofs as cisterns under *Casks and hogsheads.* ground. Thoroughly washed molasses hogsheads or wine casks, are probably the best tanks which can be used in lieu of underground cisterns. Painted outside and partly filled with water, they will shrink but little; the water will be perfectly protected from dust and foul vapors, and there will be no danger of metallic poisoning. If a single cask or hogshead is not large enough, two or more can be used, connected by pipes at the bottom. Each one should be vented separately by a small pipe. Protected externally by paint; and closed so that evaporation cannot take place, a small quantity of water will protect them from shrinkage, and save very much of the annoyance of above-ground cisterns. Water from the roof should be led to them directly, so that any shower may fill them before running into the cisterns below ground.

Washing roofs. Provision should always be made for allowing the flow from the roof to run off without entering either cisterns or tanks, because it frequently happens that after some time of dry weather, dust and dirt accumulate in the gutters, and until they have been washed out it is not best to attempt to save the water. Besides providing overflow pipes, a waste pipe taken from the very bottom of a tank is convenient, for through it much of the dirt may be drawn off and clearer water left above. Where a bottom waste pipe can be provided for large cisterns, *Waste pipes.* much dirt can be got rid of by stirring the bottom a little, and then, by opening the waste-way it flows out, while a large part of the water is saved. In some houses the tanks are placed on *Safes.* a large sheet of zinc, which has its edges turned up an inch or an inch and a half, so as to form a shallow basin. This is inclined a little, and at one corner is connected by a small pipe with the waste pipe. Any leakage is thus disposed of without going down through the house and perhaps doing damage to plastering and furniture.

The cost of setting up a tank of this sort is not necessarily *Cost of elevated cisterns.* great. An ordinary tinman can do it if a plumber is not to be had. In fact, any one who can set a pump and make the connection may set up a barrel in an upper part of the house, bring in the water to it by a pipe from the eaves and carry another pipe down to the kitchen. Connections with the wood are not *Details.* difficult, and if the tinsmith cannot achieve a wiped joint, a very good, though not very handsome one, may be made with the soldering bolt. A flange around the end of the pipe gives means for making a joint with the wood.

The size of a tank to hold the water falling upon a given roof *Capacity of cistern.* is a matter that should be carefully considered. The average rainfall in the Northern States may be taken at about 48 inches per year. That is, if all the water that falls on a given surface *Average rainfall.* in a year were saved, and none lost, we should have a depth of 4 feet. This supply is not very evenly distributed through the year. Sometimes we get 2 or 3 inches in a day, or even more, and then there are several weeks without a shower. Could we make our tank large enough to hold all the rainfall on a roof we should be very well off, but as we cannot always do this we must make some calculations as to the amount of water that falls and what we wish to use each day. The average amount of rain falling in the United States may be estimated at about 36 inches, or say 3 inches per month. For most parts of the country this will be an outside estimate, for in some months there will be no rain and in others it may reach even 6 inches.

Amount of water collected on roofs. Now, getting the length and breadth of the roof, we multiply them together and find the surface upon which we are to gather our water. This will be exact if the roof is not of too sharp a pitch. When the roof is sharp the size on the ground plan must be taken. Half the area of the roof in square feet, multiplied by 7·4 will give us the greatest amount of water in gallons which we can expect to catch in any one month, while the average will be one-half of this amount. If we have a roof of 20 feet by 40, from which we mean to take water, we shall have 800 square feet, half of this is 400, which, multiplied by 7·4, is 2960·0 gallons—the greatest amount of water we are likely to obtain, and 1480 gallons about the average quantity that we shall have for storage. By a simple calculation like this we get at the quantity of water to be expected. If in this case we find room for, say, four 63-gallon casks, we find that we shall have storage for 252 gallons, or, say, one-sixth of the water that falls. The remainder will then be available for the cistern below ground. If there was room for but one of the 63 gallon casks, we could then use about two gallons per day the year round, and never *Droughts.* run dry save in the most extreme cases. Bearing in mind the fact that there are occasional months of drought, and that our storage must, if possible, be large enough to make up for this, we can easily arrive at some idea of what amount can be used daily, and what would be needed to take us over seasons of drought. This subject is further considered in Chapter IX.

Amount of water represented by snow. In computing the value of snow, it is probably safe to assume that twelve inches of snow will make one inch of water in melting.

In the chapter on "Elementary Hydraulics Applicable to Plumbing Work," will be found many facts of interest in connection with the subject of tanks and cisterns.

CHAPTER VIII.

The Chemistry of Plumbing.

In this chapter it is my purpose to consider some facts relating to the chemical action of water on metals with which every plumber should be familiar. These facts will bear directly upon the subject of service pipes, and will include the action of water upon the metals principally used in the manufacture of pipes, tank linings, kitchen utensils, &c. As lead is the most common of the metals used in plumbing, it will, of course, receive the largest share of attention. *The field of inquiry. Its interest to plumbers. Lead.*

With regard to the safety of lead pipes for the conveyance of water, there exists a wide diversity of opinion among those claiming to be authorities in chemistry. For example, no less an authority than the Franklin Institute of the State of Pennsylvania declares, in its official publication of April, 1871, that "there exists no authenticated accounts of the health of the numerous towns and cities supplied by leaden distributory pipes having been injuriously affected." A direct contradiction to this statement had been made three months before by the Board of Health of Massachusetts, who state in their report of January, 1871, that on April 8th, 1870, they addressed to their correspondents a circular asking the following question: "Have any cases of lead colic or lead paralysis occurred in your town or district in which you have been able to trace the origin of the disease to water pipes?" To this circular 170 replies were received from as many different places. In 162 towns lead pipes were used, and of this number 41 reported affirmatively and 20 were doubtful. The affirmative cases in the 41 towns in Massachusetts were described in the report, and the record is very valuable. On the other hand, the English Scientific Commission appointed to investigate the subject *Diversity of opinion respecting lead. Franklin Institute. Mass. Board of Health. Lead poisoning in Massachusetts. English Scientific Commission.*

of the action of water on lead, made in their report the remarkable statement that "the purest water analyzed by them during an extended examination as to the most available source from which to supply the city of London, came from a lake which received all the washings from neighboring lead mines.

In recounting the history of the use of lead pipe, authorities again contradict each other. In support of the use of lead for water conduits, M. Belgrand, during some remarks before the French Academy a few years ago, alluded to the antiquity of this metal. According to Varron, the first aqueduct for conveying the Appian water to Rome was constructed in 311 B. C. From that time leaden conduits have been constantly used. The entire water service in the interior of ancient cities was of lead. Each cistern had a branch pipe which tapped a local private reservoir, which was a kind of distributing cistern for all the inhabitants of a particular section. The public fountains were supplied in the same manner. The public water service which connected the public reservoir with the private reservoir was always of lead. This mode of distribution, which necessitated very long leaden conduits for private dwellings, has been used in Paris until quite recently. It is yet employed in Rome and some other cities, and up to the end of the eighteenth century the public water service was of lead. The employment of cast iron mains only became general in 1782. There were, in fact, found in Paris a few years ago leaden conduits laid during the reign of Philip Augustus, 1180–1223. With these facts in view, Belgrand makes this decided assertion: "Until quite recently no one has seen the least danger in the use of lead. Neither Pliny, nor Frontin, nor any other ancient historian has mentioned anything resembling a case of lead poisoning. It is only during the last few years that certain persons have sought to alarm the public by showing that the use of lead conduits is dangerous." But while this eminent authority thus endeavors to plead some of the facts of history in behalf of the 900 miles of lead pipe in the city of Paris, he omits to notice others quite

THE CHEMISTRY OF PLUMBING.

damaging to his line of proof. For instance, it is not true that the dangerous character of lead was not suspected in ancient times. Vitruvius, the Roman architect (B. C. 46), forbade the use of lead for carrying water on account of its poisonous qualities, and the physician Galen (A. D. 130) condemned the use of lead as a dangerous material for water conduits. Apart, however, from the mere statements of ancient authors, M. Belgrand presents some remarkable cases of the preservation of lead pipes. He states that portions of the public water service are taken up from time to time, and that "their interior surface is always smooth and shows no traces of corrosion." He submitted for the examination of the Academy two pieces of lead pipe, one taken from a conduit in the Faubourg St. Antoine, which was laid in 1670. After this period of more than 200 years, the impressions of the sand of the mould in which the pipe was cast were distinctly visible in the metal. The other sample was not quite so old. *[margin: Contradictory testimony. Vitruvius. Galen. Remarkable life of lead pipe in Paris.]*

In seeking for an explanation of facts apparently so contradictory, we are met by a variety of results which, viewed without their connections, seem to leave the problem unsolved. In estimating the value of such results, however, it must be borne in mind that in the laboratory we can only operate on small portions of lead and small quantities of water. We also find it difficult to place the two substances in precisely the same conditions as they exist in practice. Waters which in the laboratory seem to corrode the lead very slightly, often act very violently on the pipes through which they are conveyed. Here, then, is a very important source of error in our investigations. The subject, therefore, abounds in uncertainties. The chemist can state as a general fact that water, under some circumstances, will act on lead, but it is difficult to tell in advance how much lead a given water will dissolve. The physician is also met by a still more puzzling fact, viz., that different persons are differently affected by given quantities of lead. *[margin: Danger of mistaken conclusions from isolated facts. Laboratory tests. Uncertainty of chemical investigations. Susceptibility to lead poisoning.]*

150 THE CHEMISTRY OF PLUMBING.

Physical properties of lead. The physical properties of lead are such as to peculiarly adapt it for use as a material for water pipes—so far, at least, as convenience of manipulation and cheapness are concerned. These are so well known that it is unnecessary to discuss them here. **Chemical composition.** It is important, however, to know its chemical composition, as we shall have occasion to consider some of the impurities contained in the ordinary lead of commerce. It alloys **Impurities in the lead of common use.** very readily with tin, bismuth and antimony. In its crude state it usually contains some of its own oxide mechanically mixed with it. This impairs its malleability and ductility, but increases its resistance to pressure. Commercial lead usually contains also metallic impurities. Hard lead owes its hardness **Reich's analyses.** to the presence of antimony. The following analyses of various kinds of lead, by Dr. Reich, are trustworthy:

Metals.	German Lead.		Hard Lead.	Antimonial Lead.	
	Raw.	Refined.		First Specimen.	Second Specimen.
Lead..................	97.92	99.28	87.60	90.76	87.60
Arsenic...............	1.36	0.16	7.90	1.28	0.40
Antimony..............	0.72	traces	2.80	7.31	11.60
Iron..................	0.07	0.05	traces	0.13	traces
Copper................	0.25	0.25	0.40	0.35	traces
Silver................	0.49	0.53	....	....	....

Water corrodes lead in proportion to its purity. It is a remarkable fact that water corrodes lead with a vigor proportioned to its purity. This was the theory of Sir Robert Christison, one of the early experimenters in this department of chemistry, and it has been sustained by more recent investi-**English experiments.** gations. In experiments made by the English commission appointed to investigate this subject, it was found that pure water in contact with lead for 24 hours became highly poisonous.

M. Besnon on the corrosion of lead. A very clear explanation of the action of water on lead was presented before the French Academy of Sciences by M. Besnon. He says: "Rain water and distilled water attack lead recently cut with great rapidity, and form upon the surface and in the water a white, partially crystalline precipitate. The steam formed in the distillation of fresh or salt water attacks, in condensing, the coolers composed of an alloy of tin and lead. On the first day of the operation a greater quantity of lead is taken from the surface than on following days." For example, in M. Besnon's experiments the quantity of lead discovered in the water after the first day's distillation amounted to 2·17 grains to the gallon. This proportion was decreased to 1·82 grains per gallon on the third day. In the distillation of water for pharmaceutical purposes the same thing takes place. Consequently, such preparations as the extract of orange flowers, brandy, &c., contain traces of lead.

The action of pure water on lead has been regarded as a subject of great importance by the French experimenters, and the results of their experiments are very interesting and conclusive. M. La Pierre states that he passed a current of steam through a horizontal lead coil, which was conducted into a reservoir containing water. The admission of the steam and the amount of the water in the reservoir were regulated in such a manner that only a portion of the steam was condensed, while the rest escaped to the atmosphere. To cleanse the coil, steam was passed through it for eight hours. Steam was then allowed to circulate slowly through the coil for three days. The water obtained by condensation had a cloudy and milky appearance, and gave a deposit of the carbonate of lead. The water so collected was filtered, and the residue amounted to 5·22 grains per gallon of water. The water, after filtering, did not give very satisfactory indications of lead, showing that the corrosion of lead by distilled water gives an insoluble powder—a very important fact. M. La Pierre treated this filtered water with carbonate of ammonia and then filtered again. The latter reagent gave a very slight precipitate of lead, which was left on the filter paper. The water coming from the second filtration was evaporated, and even this gave a residue of lead amounting to one-quarter grain to the gallon.

Action of pure water on lead.

M. La Pierre's experiments.

THE CHEMISTRY OF PLUMBING.

Impurities in distilled water. It has been argued that such experiments do not prove that absolutely pure water does not attack lead, but that the water obtained by distillation, not being perfectly pure, acts violently by reason of the presence of certain salts. For instance, ordinary distilled water contains traces of nitric or nitrous acids, in combination with ammonia, forming nitrite and nitrate of ammonia. These salts are volatile, and pass over with the steam during the process of distillation. They act vigorously on lead, forming nitrite and nitrate of lead, which, it is claimed, are converted almost immediately by the carbonic acid of the air into carbonate of lead. This is not of great importance, for as water obtained by ordinary distillation is as pure as any found in nature or used in ordinary operations, we are not interested in water purer than this. **Dr. Christison's experiments.** Dr. Christison seems to have settled this question very satisfactorily. He added to a certain quantity of water some potash, which combines with nitric and nitrous acids, forming compounds not volatile. He then distilled the water, taking great care to prevent the access of impurities from other sources. He subjected some lead to the action of water thus obtained, and found that its action was even more vigorous than in the case of the ordinary distilled water of the laboratory.

Air in water The action of air in facilitating the corrosion of lead by water, is probably very important. If a sheet of lead with a bright surface be immersed in a vessel of freshly boiled distilled water, and the vessel be tightly closed so that no air can obtain access to it, the metal will remain for a long time with scarcely any symptoms of tarnishing. But if the vessel is uncovered, the water immediately begins to absorb air, and soon the luster of the metal is dimmed and small scales of oxide of lead begin **Corrosion at the line of partial submersion.** to form upon it. If a sheet of lead is partially submerged in water in an open vessel, there will be found upon it after some days, at the line where the lead meets the surface of the water, yellowish white crystals of hydrate of lead, with crystals of the carbonate of lead also. This line, of course, marks the place where both air and water act on the lead at once. Again, if a piece of lead with a bright surface be exposed to a dry atmosphere, and another to a moist atmosphere, it will be found that the latter has become tarnished by the formation of a bluish-gray coating, while the former remains bright.

<small>Atmospheric influences.</small>

There is no lack of facts to prove that air is an effective auxiliary to water in its action on lead. These facts, while demonstrating the theory, no less prove its practical importance. M. Bobierre remarks that "when the lead sheathing of a vessel is corroded and perforated by the action of water, the damage is principally sustained at the water line; that is, where the sheathing is alternately subjected to the action of water and the oxygen and carbonic acid of the air. Those portions of sheathing that are always below the water line are, on the contrary, far less acted upon." In a paper presented by the same writer to the French Academy of Sciences, he mentions a very interesting and important illustration of the combined action of air and water. His attention was called on one occasion to a leaden reservoir used in a hydropathic institution, which had been eaten through with holes. The corrosion had taken place quite rapidly, although the lead was of the finest quality. A careful examination of the case showed, also, that there was no fault in the workmanship of the plumber. It appeared, however, that the tank was often entirely empty and was subsequently filled by a stream of water from a cock 3 feet above that portion of the surface which the falling liquid struck. Here was an excellent opportunity for the combined action of water and air, and the result showed the effectiveness of this combination. A carbonate of lead was probably formed upon the metal, and when the cock was turned on this coating was washed away from the one spot by the falling water, allowing an opportunity for the formation of a fresh coat in the place from which the former crust was removed. M. Bobierre also states that on being called in a very aggravated case of lead poisoning in Nantes, he discovered

<small>Air and water</small>

<small>M. Bobierre.</small>

<small>Corrosion at a fluctuating water line.</small>

<small>Examples.</small>

that the lead pipes were coated internally with a muddy crust of carbonate of lead. He found, also, that the pipe, by reason of its position and numerous curves, formed in certain places air traps, thus offering all the facilities for an active oxidation. At Nantes such a case of poisoning is quite uncommon, although the pipes are of lead. The reason for such immunity M. Bobierre finds in the fact that the metal is kept constantly in contact with water. In support of this theory, he presented to the French Academy the results of a very interesting experiment. He placed two lead pipes in contact with distilled water under different conditions. One was completely immersed, while the other was laid so that half should be in the water and half out of it. He also placed on a porcelain plate a little conical heap of crystalline fragments of lead, and then added water to a depth equal to half the hight of the metal. After eight hours the water in the vessel containing the tube completely submerged, was found to be but little affected by the metal. A very marked reaction was obtained in the case of the water which only half covered the tube. In the case of the water surrounding the broken fragments, the action of the air in this case being evidently the greatest, the corrosion had been so vigorous that the water actually appeared milky from the carbonate of lead held by it in suspension. It is a question, then, of considerable importance whether the lead is exposed to the alternate action of water and air, or is always in contact with water. "What is true of a pipe," says Bobierre, in speaking of the action of water on lead, "would be true of a tank; and a pipe which is always kept full by the pressure of the water in a reservoir above it, is less liable to corrosion than a pipe which is filled by a pump and in which the water does not remain for any length of time."

<small>Bobierre's experiments and conclusions.</small>

<small>Moisture and carbonic acid.</small> It is known that when lead is exposed to the action of moisture and air highly charged with carbonic acid, it is readily <small>Bloxam.</small> attacked. Bloxam states that the lead of old coffins is often found converted into a white, earthy-looking, brittle mass of

basic carbonate, with a very thin film of metallic lead inside of it.

The corrosion of lead pipes is often facilitated by heat. In the latter part of 1869, Dr. Wallace, in an address before the Glasgow Philosophical Society, remarked that a new and very different source of danger was revealed by the analysis of water taken from the cisterns in various parts of Glasgow. It was found that water which had become warm by remaining in pipes that were exposed to heat, either by proximity to the usual hot-water pipes or otherwise, was frequently contaminated with lead to such an extent "as to render the use of the water for dietetic purposes dangerous." Some time ago we had a painful illustration of this new source of danger in a case of lead poisoning from the use of Croton water. An elderly gentleman in this city was completely prostrated by paralysis. His physician, judging that his symptoms indicated lead poisoning, investigated the matter and discovered that the patient had been using wheaten grits for dyspepsia, and that the cook was accustomed early every morning to soak them before boiling. She used for this purpose the water which had stood in the hot-water pipes over night. It is probable that water might have been drawn from the cold-water pipes without as serious mischief. The practice is, however, one which all housekeepers should forbid, and the rule of the house should be to allow the water which has stood for several hours in a lead pipe to be run off before any is drawn for drinking or for use in the preparation of food.

Prof. W. R. Nichols presents, as an explanation of the increased corrosive energy which heat imparts to water, the suggestion that the alternate contractions and expansions of the pipe produced by changes of temperature cause a disarrangement in the particles of the lead and a change in its mechanical structure. A physical action, according to this theory, develops or accelerates the chemical forces which previously are either dormant or feeble. Whatever may be the reason, it is important to be sure of the fact, and the following results of an investigation by the same professor give strong evidence of the possibility of a dangerous energy being awakened in water by the agency of heat. A sample of water was taken from the pipes of a private residence in Boston. The water analyzed had flowed through 100 feet of tin-lined pipe and then through 10 or 12 feet of lead pipe. The pipes had previously been in use for six months. The sample gave ·029 grain of metallic lead to the gallon, or only ·0342 part in 100,000. A sample was then taken in the same residence after the water had flowed through 40 additional feet of lead pipe (hot-water pipes), through a lead-lined tank and an ordinary copper boiler. This sample, on analysis, gave ·112 grain to the gallon, or ·191 part in 100,000. The latter results from water which had flowed through only 50 feet of lead pipe (40 feet being hot), compare very strikingly with the results obtained in the chemical laboratory of the Massachusetts Institute of Technology. In this case the water had flowed through 150 feet of cold lead pipe and had stood in the pipes for 14 hours. The water then only yielded ·057 grain to the gallon, and after enough water had been removed to clear the pipes, there was found in the sample taken only ·0179 grain to the gallon. The results in the case of the hot-water pipes were, therefore, not due simply to the length of pipes.

A most remarkable and conclusive case was disclosed by Prof. Roscoe, of Manchester, England, in November, 1874. He states that he had received from a surgeon of Manchester a white powder taken from the inside of the covering of a leaden hot-water cistern. The interior of the cover presented a honey-combed surface, and in many places stalactite masses hung down which were from one-quarter to one-half of an inch in length. Manchester water has been found, after a considerable experience of its qualities, to be, when cold and under ordinary circumstances, quite harmless as regards lead. In this case, however, an extraordinary activity had been developed. An analysis of the powder revealed it as a hydrocarbonate of lead.

Lead oxide (PbO).................... parts.   85·67
Carbonic acid.......................   "   12·12
Water..............................   "   2·21

Total............................... 100·00

Prof. Roscoe presents a very plausible theory for this action. *Roscoe's explanation.* He thinks that the powder was formed by the action of water which had arisen as steam from the surface of the water in the cistern and condensed on the cover. This condensed water would be very pure and would contain air in solution, which would act on the lead.

A simple chemical fact, usually but little considered, adds *Chemical basis of empirical proof.* weight to the empirical proof presented above. I have already described some experiments made to determine the action on lead of the hot vapors formed by the distillation of sea water. These usually contain some compound of magnesium, generally *Mineral salts.* the chloride, but often the iodide, bromide and sulphide, and the corrosive action of these salts on lead is always increased when assisted by air and when their temperature is raised to the boiling point of water. Of all the inorganic acids, nitric (*aqua for-* *Inorganic acids.* *tis*) is the only one of common occurrence that acts on lead under ordinary circumstances. Of the organic acids commonly used, *Organic acids.* only acetic acid corrodes lead. The latter acid is the active principle of vinegar. Sulphuric acid (oil of vitriol), unless highly *Oil of vitriol.* concentrated, has so little effect on lead that stills of this metal are used in the first concentration of the acid. Indeed, a little sulphuric acid present in water acts as a good preventive against the corrosion of the lead. Carbonic acid, as we shall see here- *Carbonic acid* after, acts on a moist surface of lead; but I use the term acid in this connection in its popular acceptation. Nitric acid does not *Nitric acid.* often come in contact with lead, but the fact that acetic acid *Acetic acid.* corrodes the metal is of importance. Lead faucets have fre- *Vinegar faucets.* quently been used for vinegar barrels, and as good vinegar should contain from $3\frac{1}{2}$ to 4 per cent. of acetic acid, there is great danger in this practice. Such vinegar unquestionably contains lead, and in the form of a virulent poison known as the acetate of lead (sugar of lead). Prof. Chas. E. Munroe, of Harvard University, analyzed two samples of vinegar that had been drawn from a cask with a leaden faucet. One sample contained 25 grains of sugar of lead to the quart and the other 144 grains of sugar of lead to the quart. Prof. Hill, in reporting this fact, mildly characterizes the sale of such poison as "criminal carelessness." Our language affords somewhat stronger epithets appropriate to the case.

Lead is found sometimes in cider and other liquors. It is due in these cases to the use of lead in certain processes of their manufacture. The fermentation which takes place when the juice of apples is converted into cider results in the production of organic acids, and if the process is carried on sufficiently long, cider vinegar containing acetic acid is the product. A case of poisoning by lead contained in cider is recorded by Dr. Horatio Adams, of Waltham, Mass. The person so afflicted exhibited, after drinking the cider, the usual symptoms of lead colic, which were afterward followed by a partial paralysis of his extremities.

The action of organic acids on lead is much more active than that of inorganic acids. The distinction between the two terms, organic and inorganic, so far as their meaning concerns us in this discussion, is simple. We find in lake and river water various acids derived from the decay of bodies once organized. Such acids are called organic, whether they are derived from the decay of animal or vegetable organisms, the latter, however, being the chief source. Vinegar (acetic acid) is an illustration, since it is formed by allowing certain bodies, as wine and apple juice, to ferment or decay (fermentation being an operation that goes on in the process of decay). Organic acids of various constitution are formed in natural waters, but they are so transient in their character that before we can separate them from the water in which they exist they have changed their form. As the decay of wine produces acetic acid, so the decay of wood produces formic acid, &c. These acids frequently decompose soon after their formation by contact with other bodies, and change their constitution. Sometimes they are soluble in water and color it; at other times they are insoluble and are simply suspended in it. These vegetable acids are very powerful in their action on lead. They induce the formation of oxide of lead, which is equivalent to rusting the lead, and then dissolve the oxide (or rust) so formed. Dr. Samuel L. Dana, of Lowell, Mass., remarks that he has found vegetable acids so abundant in the Merrimac River as to dissolve in 24 hours as much lead as pure water. The vegetable acids also act upon certain insoluble salts formed on the lead by the action of other materials dissolved in the water, and cause them to become soluble. Of this and of the general subject of the action of organic matter on lead, I shall have more to say further on. *Character of organic acids.* *Their action on lead.* *Dr. Dana.*

Potash and soda, if they exist in water in very small quantities, do not act continuously on lead. They corrode it very energetically at first, causing the formation on the metal of an oxide of lead, but the latter is soon converted by the carbonic acid of the air dissolved in the water into carbonate of lead. This forms an insoluble coating over the interior surface of the pipe and defends the latter from corrosion. If the alkalies exist in greater quantities in the water, another effect is produced. *Potash and soda.*

Water which has been affected with organic matter, as is the case when it comes from or flows in the vicinity of vaults or cesspools, is often very strongly alkaline. The alkali in this case is neither potash nor soda, but ammonia in various combinations. Such waters have a destructive influence on lead and operate in a very indirect manner. I have explained that the oxygen of the air acts on the lead, forming an oxide which is somewhat soluble and very poisonous. The carbonic acid, a gas which always exists in the atmosphere and is generally dissolved in water, usually combines with this oxide of lead and *Organic impurities in water.* *Carbonate of lead.*

forms a carbonate of lead. The latter operation is attended with two good results: First, the carbonate of lead is insoluble and therefore not to be feared as a poison; second, it is formed as coating or crust on the interior surface of the lead and pre-*Dissolved by alkalies.* vents the further action of the water on the metal. When, however, the water contains potash, soda, ammonia or lime, these alkalies combine with the carbonic acid and prevent its action on the oxide of lead formed on the pipe by the action of the water. The consequence is that the poisonous oxide of lead, instead of being carbonated, is carried off by the water and a fresh, unprotected surface of lead is left to be corroded.

Indirect action of alkalies in water. The alkalies, as I shall explain, have a further injurious effect in aiding the water to dissolve salts of lead formed by certain waters, and which, being insoluble, would otherwise protect the metal. The action is indirect but very corrosive.

Lime. Lime acts in both these ways with great energy, and although the transparency of the water may remain unimpaired, the tests for lead will reveal it in large quantities. It is scarcely prudent, then, to join lead pipes with cement. A number of cases have, in fact, occurred of corrosion of lead tanks or cisterns by pieces of mortar that have dropped into them, the lime of which causes the oxidation of the metal. Sometimes, too, a lead pipe is laid in or conducted through fresh mortar, which is *External corrosion of lead pipes in mortar.* frequently moistened. Here an outside corrosion goes on with great vigor. Considerable quantities of fresh mortar are frequently deposited in pipes during the erection of buildings. The open ends of such pipes should, therefore, be carefully closed during the period when this is liable to occur.

Natural waters not pure. Water as we usually find it in streams, wells and springs, is seldom even approximately pure. We have considered the manner in which pure water acts on lead, and it is necessary now to explain the manner in which the presence of foreign substances in water varies this action. The action, of course, is to a great degree dependent on the nature of the substance it contains in solution. It is important, therefore, as well as interesting, to

THE CHEMISTRY OF PLUMBING.

recall for a moment the varied composition of water as found in different parts of the world.

Water is an almost universal solvent. As it falls through the air it dissolves the gases that are in the atmosphere. Rain water is the nearest approach we have in nature to pure water, but it generally contains ammonia and carbonic acid, and has been found to hold in solution even sulphate of lime and organic matter of animal and vegetable origin. In Paris it has been found to contain traces of iodine and phosphoric acid. Rain water collected near the sea always shows traces of chlorides. In cities and their vicinity, rain water is usually more impure than in other places. The first rain that falls during a storm is more impure than that which falls afterward, since the first washes the impurities out of the air. Dr. Dana states that after long-continued observation and analysis he has become assured that nearly all the salts of the ocean are found in rain water—of course in minute quantity. The amount of solid matter in rain water he estimates at 1·603 grain per gallon. This is a very small quantity, but from this basis he estimates that one inch of rainfall yields about one grain of solid matter per square foot, which is equal to 6·268 pounds per acre. These figures mount up rapidly when he states that at the rate of 30 inches annual rainfall on the district included in the 12,077 acres of Lake Cochituate and its drainage land, there are deposited 2,270,959 pounds salt. Thus the trifling quantity of solid matter in the rain becomes very great when estimated in the aggregate. As soon as the rain reaches the earth it begins its work of dissolving away portions of the soil, and thereafter, until it reaches the sea—where the water attains its greatest density—it keeps increasing in impurity. The character of the water, then, is determined by the soil over which it has flowed. The purest natural water is generally found in deep lakes and in slaty and granitic districts, the material of such formations being very insoluble. The water supplied to the city of Glasgow is brought from Loch Katrine, which contains only two grains of solid

Properties and composition of potable waters.
Rain water.
Ocean salts in rain.
Solid matter.
Lake Cochituate.
Character of water determined by soil.
Loch Katrine.

162   THE CHEMISTRY OF PLUMBING.

Croton. matter to the gallon. This compares very favorably with the Croton water of New York, which contains 6·66 grains to the *Thames* gallon, or the Thames, of England, which contains 20 grains to the gallon. The solid matter dissolved in river water, such as is used for drinking, varies in weight from 6 to 50 grains.

Substances found in river water. Among the substances, some of which are familiarly known, found in solution in the water of rivers, are sulphate of soda (Glauber's salt), sulphate of lime (gypsum), sulphate of magnesia (Epsom salt), sulphate of potash, sulphate of alumina, chloride of sodium (common salt), chloride of potassium, carbonate of lime (chalk), carbonate of magnesia (common magnesia), carbonate of iron, carbonic acid, silica (sand), sulphuretted hydrogen (the peculiar ill-odored gas found in sulphur waters), phosphate of lime (a substance found in bones), bromides and iodides of calcium and magnesium, and vegetable and animal sub-
Impurities derived from towns. stances. Near large towns, the water frequently contains salts of nitric and nitrous acids (nitrates and nitrites) and ammonia.

Spring and well water. Spring and shallow well waters generally contain plenty of sulphate and carbonate of lime, and but little salt. The solid matter held in solution often amounts to 150 grains per gallon.

Artesian wells. Deep or artesian wells generally contain from 50 to 70 grains of solid matter per gallon. They mostly contain a remarkable proportion of soda salts, and generally carbonic acid. They usually have but little organic matter or lime, but often hold a considerable quantity of the phosphates in solution.

Unusual ingredients in water. If water did not sometimes contain unusual ingredients in solution, and if no disturbing local causes ever operated, the much-discussed question of the safety of lead pipes would be a very simple one. But, unfortunately, water often contains matters in solution which change the order of chemical action already described, and cause it to act quite rapidly on the metal, often destroying it altogether. On the other hand, the presence of some foreign substances in solution may assist in the pro-
Most waters incapable of corrosive action on lead. tection which some waters normally exercise on lead. Hundreds of analyses have been made of water from springs, wells, ponds, lakes and open reservoirs, and the results of these examinations

THE CHEMISTRY OF PLUMBING. 163

clearly demonstrate that most waters contain substances which render safe the use of lead pipes. Even rain water, which in its pure state acts on lead, becomes sufficiently impure by flowing over roofs and being collected through gutters, especially in large towns, to be incapable of vigorous action. Unless their action is interfered with by the presence of other substances in sufficient proportions, the carbonates, sulphates, phosphates and borates exercise a protective influence over lead. On the contrary, the sulphurets or sulphides, nitrates and nitrites, chlorides, organic matter and the alkalies impart to water a corrosive power over lead. *Protective impurities. Corrosive impurities*

Well water usually contains a considerable quantity of nitrates and chlorides, not only in this country, but in Europe. Spring, pond and river waters do not contain so much. As might be expected, well water is often corrosive to lead, the nitrates having this action. Dr. Dana says in all cases which he has examined where lead pipe has been introduced into the well for the purpose of a conduit, the water has been found to contain lead years after the pipe was first used. Such pipes have been eroded deeply, and in many instances perforated and thereby rendered useless. Lead pipes used in wells have, in some cases, been perforated in six months. In one case in Lowell where a new well was sunk in a district not previously crowded with inhabitants or exposed to the drainage of the locality, the lead pipe conducting the water was so eroded as, at the end of three years, to become entirely useless. New pipe was supplied which also became useless at the end of about the same time, and the third set of pipe (in use at the time of writing) requires frequent repair, having performed nearly the usual service of its predecessors. It is not uncommon for the erosion to go on within the pipe until the film of lead forming the outer boundary of a pit (or cavity on the interior of the pipe) is too thin longer to sustain the atmospheric pressure. When the pump is first partially exhausted of air, the film gives way, bursting inwardly. The plumber having mended one *Nitrates and chlorides. Corrosion of lead by well water. Investigations in Massachusetts.*

hole, produces another in the attempt to "catch the pump." These facts are not peculiar to the well water of Lowell. Long ago Dr. Wall, of Worcester, England, noticed that a lead pipe was destroyed in about three years by water at a farm house. Dr. Hayes, State Assayer of Massachusetts, confirms these facts by similar observations on lead pipes in the wells of Boston and its vicinity. In one case under his observation, the portion of the pipe immersed was corroded so completely that it separated and fell to the bottom of the well. Cases of severe lead colic, paralysis and neuralgic affections, have occurred in instances when well water has been conveyed through lead. Aside from the abundance of nitrates in such waters, one of the reasons for the superior activity of well water on the lead is the greater quantity of carbonic acid from the air held in solution in such water. This is on account of its coldness. Gases dissolve more readily in cold than in warm water. When well water is brought to the surface and exposed to the temperature of the outer air, it parts with a portion of its carbonic acid and becomes flat and insipid. Spring water not containing so much of the nitrates and the chlorides, is less energetic in its action on lead. Cases of corrosion do occasionally occur, however. An instance is related of a minister residing at Dedham, Mass., who was attacked by a complication of affections which forced him to go traveling for his health. He returned well and, unsuspicious of the cause of his troubles, began again the use of the same water which he had been drinking before he left. He was again prostrated, and this time it was supposed he would lose his life. The water was examined and found to contain lead. It had been conveyed in a lead pipe for half a mile from a spring. The water was abandoned and the disease disappeared.

The chemical law by which the dividing line is drawn between the two classes of salts, those on the one side protecting the lead and those on the other promoting corrosion, is simple, but it can scarcely be explained to those unfamiliar with chemistry without being somewhat elementary and explaining the

THE CHEMISTRY OF PLUMBING. 165

meaning of the terms carbonates, sulphates, &c. This I shall endeavor to do as briefly as possible.

Among the range of chemical compounds found in nature or produced in art, is a class of bodies called acids. Many of them are familiarly known in the practice of the mechanical arts. Sulphuric acid, commonly called oil of vitriol, is an instance. It is very corrosive and will combine with most of the metals quite readily. When sulphuric acid thus combines with a metal, the combination is termed a sulphate of that metal. Thus sulphuric acid combines with zinc and forms a sulphate of zinc (white vitriol), a substance much used in calico printing, and as a medicine. Sulphuric acid also combines with lime and forms sulphate of lime (gypsum), a substance often contained in solution in water. It must be understood, however, that although sulphuric acid is a very corrosive acid, it loses this property when it combines with anything, such as lime. Nitric acid is another illustration of this class of bodies. It is also corrosive, and is known in commerce as *aqua fortis*. When nitric acid combines with anything—for example, lime or potash—we call the combination nitrate. In the cases mentioned, we say nitrate of lime or nitrate of potash. The combination has lost the properties of the acid, for when an acid combines with a base, as the potash or lime would be called by chemists, each neutralizes the other and the compound has entirely new properties. Carbonic acid forms carbonates when it combines with a base. For instance, chalk is a combination of carbonic acid and lime—a carbonate of lime. Phosphoric acid, which is formed whenever the phosphorus end of a lucifer match burns, forms phosphates. There exists in the human bones a substance called phosphate of lime. It is simply a compound of phosphoric acid and lime. I have already mentioned the borates as a class of substances that exercise a protective influence on lead. Borax is an instance of a borate; chemically it is known as borate of soda. The termination "ate" indicates that some acid has combined with something

[Marginal notes: Acids. Sulphuric acid. Sulphates. Nitric acid. Nitrates. Carbonic acid. Carbonates. Phosphoric acid. Phosphates. Borax. The termination "ate."]

else called a base. Such terms as chloride and sulphide also mean that an acid has combined with a base, but the termination "ide" shows that it is another kind of acid that has entered into combination. Chloride of iron would mean that an acid known as hydrochloric acid has combined with iron. The termination "ide" instead of "ate" is only a device to denote what kind of an acid it is that has entered into combination. For instance, chlorate of iron would mean chloric acid and iron, while chloride of iron would mean hydrochloric acid and iron.

It is a peculiar fact that when an acid has entered into combination with a base, if another base presents itself the acid will often give up the first base and attach itself to the new base. For instance, water sometimes contains sulphate of lime in solution. In passing through a lead pipe, the lead offers a superior attraction for the sulphuric acid contained in the sulphate and the acid leaves the lime and goes to the lead, forming a sulphate of lead. So, also, a phosphate of lime in passing in solution through a lead pipe, will become decomposed, the phosphoric acid going to the lead and forming a phosphate of lead. And so it is with the other "ates" and "ites" and "ides;" they are apt to form a carbonate, a phosphate, a borate or a nitrate of lead, or a chloride or a sulphide of lead, or a nitrite of lead. These new combinations form on the inside of the lead pipe as a crust. Now, if the new compounds are themselves insoluble in water, they will generally continue to form on the inside of the pipe until they have become sufficiently thick to protect the pipe from the action of the water. If, on the contrary, they are very soluble in water they will be washed away, and a fresh surface of lead being exposed the corrosion will go on until the metal is eaten through. This explanation will render clear the proposition which Dr. Christison has stated, that "the proportion of each salt" (for instance, the sulphate of lime, since all such compounds are known in chemistry as salts) "required to prevent

THE CHEMISTRY OF PLUMBING. 167

action, is nearly in the inverse ratio of the solubility of the compound which its acid forms with the oxide of lead." Theoretically, then, the answer to the whole question would seem to depend on the solubility of the salts of lead. The carbonate of lead is soluble in 50,000 times its own weight of water. This means that it is practically insoluble. The sulphate of lead is soluble 20,000 times its own weight of water. So far as chemists have been able to tell, the phosphate of lead is altogether insoluble. The nitrate of lead, on the other hand, is soluble in only three parts of water—that is to say, a gallon of water has the capacity of dissolving one-third of its weight of nitrate of lead. Chloride of lead is soluble in 135 parts of water (a dangerous degree of solubility). The oxide of lead is dissolved by pure water at the rate of 8 grains per gallon. {Solubility of lead salts.}

The problem of telling in advance whether a particular sample of water will act favorably or unfavorably on lead, would not be so difficult if the theory just developed was allowed to remain undisturbed in its application. The trouble is, however, that other laws are at work distributing and complicating our calculations. For instance, Dr. Nevins has asserted that the salts above described as protective of lead, are only so when they are present in small proportion. When present in large proportions, they seem to permit an action upon the lead. Dr. Nevins claims that he has obtained results in his experiments which warrant him in this statement. It may be, perhaps, that the salt of lead formed on the pipe, while not soluble in ordinary water, is soluble in water containing an excess of the substance which caused the deposition of the crust on the lead. The following is an illustration of this peculiar chemical law. If we mix some lime in water and, after allowing the lime to settle, pour off the water, we will have a perfectly transparent liquid containing lime in solution. If we now direct a stream of carbonic acid gas through this liquid by means of a tube, it will become quite milky, and in the course of a short time a white powder will settle at the bottom. A chemical change has been {Theoretical estimates of the corrosive action of water on lead.} {Dr. Nevins.} {Salts of lead dissolved by the substances forming them.} {Experiment with lime.}

effected in the liquid. The carbonic acid has united with the soluble lime and formed a carbonate of lime which is insoluble in water and therefore sinks to the bottom. If we then stir up the mixture and continue to drive a stream of carbonic acid through the liquid, the latter becomes clear again. Another chemical change has been accomplished. The carbonic acid having changed all the lime into carbonate of lime, went on dissolving in the water, and soon the liquid became a strong solution of carbonic acid. Now while carbonate of lime is insoluble in water, it is soluble in a solution of carbonic acid, and therefore dissolves and disappears. Whatever may be the reason for the action that Dr. Nevins describes, it is well to bear the fact in mind. In relation to it Dr. Christison remarks, that if the protective salts are not protective beyond a certain limit, it is necessary to fix that limit before we can deduce any practical results from the suggestion. Dr. Nevins has not done so.' From all that we know of the constitution of natural waters that are applied to household use, it is more than probable that the proportion of salts necessary to change their own action from that of protection to corrosion, is greater than is ever likely to occur outside of the laboratory.

Protective salts protective only to a limited degree.

The most perplexing question in connection with the subject is that which regards the mixture of salts, some of which are protective and others corrosive. Thus, suppose nitrate of lime (a corrosive salt) and carbonate of lime (a protective salt) are found in the same water, what will be their probable action? The answer, no doubt, depends on the proportion of the two substances as they exist in the water. This matter, however, can better be explained after we have considered the action of each class of substances which occur in water. But after all the questions depending for their answer on theory have been disposed of, there still remains the fact that local causes may give us very unexpected results. Prof. Nichols speaks of the case of Dr. Treadwell, of Salem, Mass., who suspected that he was suffering from lead poisoning, and who sent to him for

Action of mixed salts.

Lead poisoning in Salem.

THE CHEMISTRY OF PLUMBING   169

analysis samples of the water supplied to his house. Lead was found to be present in the water in large quantities. A specimen of the water from the same aqueduct, but taken from another locality, afforded only a trace of lead. Here some local cause was operating. Dr. Christison, whose observations on this subject are always of great value, remarks that unforeseen circumstances may counteract all the preservative effects of any particular water. <small>Influence of local causes.</small>

Most waters, fortunately, contain carbonate of lime, and this substance is the most effective protector of lead that exists in water. It is to the presence of this salt in drinking water that we owe the absence of lead in most cases in which the test is made. I have already explained how the carbonic acid of the air dissolved in the water combines with the oxide of lead, and thus reaches the latter harmless. It is in a slightly different way that the carbonate of lime dissolved in the water affords the same protection. Carbonate of lime, which is seen in nature as chalk, limestone and marble, is not soluble in water— or at least is practically insoluble, one part requiring for its solution more than 10,000 parts of water. But as it was shown in the little experiment referred to above, the carbonate of lime is soluble in water containing carbonic acid in solution. If some carbonate of lime be dissolved in water containing carbonic acid gas in solution, and the latter be removed by boiling, the water will no longer hold the carbonate of lime in solution; the particles of the carbonate will soon be seen falling to the bottom, giving the liquid a milky appearance. This experiment can be easily performed by any one, by simply boiling a little calcareous (or limestone) water. What occurs in this experiment is similar to what occurs in the lead pipes, only the carbonic acid is withdrawn from the water in another manner. The oxide of lead is formed by the action of the water on the pipe, as has already been explained, and this oxide of lead combines with the carbonic acid dissolved in the water, as previously shown. The carbonate of lead is formed <small>Carbonate of lime. Chalk, limestone and marble. Action of carbonate of lime on lead</small>

and, being insoluble in the water, collects on the pipes. But the carbonic acid, having been removed from the water by the oxide of lead, the water can no longer hold the carbonate of lime in solution, and this collects on the pipes also. Consequently, the crust formed on the inside of lead pipes in districts in which the water contains any limestone, is composed of a mixture of carbonate of lime and carbonate of lead, both of which being insoluble soon become of sufficient thickness to defend the pipe from the action of the water.

Carbonates of magnesia and iron. Other carbonates are present in water besides that of lime. In discussing the constitution of water, I referred to carbonates of magnesia (common magnesia) and carbonate of iron. The former, though not very soluble in water, is much more so than carbonate of lime. It is very soluble in water containing carbonic acid. It acts in a similar manner to the lime carbonate, and protects the lead against the corrosive action of water.

M. Dumas' experiments. An experiment noted by the French chemist M. Dumas is interesting, not only as tending to prove the statements already made regarding calcareous salts, but as being one which any person with moderate skill can try for himself. He took five bottles, and placed in each some pellets of lead. He then poured into the first some distilled water, into the second some rain water, into the third some water from the Seine, into the fourth. water from the Ourcq (the drinking water of Paris), and into the fifth some well water. He allowed them to stand, and presently tested them with sulphuretted hydrogen, a delicate test for lead. The distilled water gave unmistakable signs of lead. The water in the other bottles showed no lead whatever. All of the latter specimens contained calcareous salts.

Carbonate of soda. It is claimed that there are two or three substances whose presence in water prevents the carbonates from exercising their protective influence. One of these is carbonate of soda (soda ash), but so far it has never been shown how much of it there must be in the water to interfere with the action of the other carbonates; and besides, as this substance always exists in water as the bicarbonate, which has no such effect as is claimed for the carbonate, the question seems of little importance. Another *Carbonic acid gas.* substance is carbonic acid gas itself. We have seen that the carbonate of lime is dissolved in water containing carbonic acid in excess. Now, unfortunately, if the carbonic acid is in great excess, it enables the water to dissolve some of the lead. Herein is the great danger of employing lead pipes in soda- *Lead pipes in soda-water fountains.* water fountains, for this beverage, being nothing more than water highly charged with carbonic acid, acts vigorously on the pipes and becomes poisonous. Fortunately, however, in nature waters highly charged with carbonic acid are rare and are generally medicinal in their character. Such are the waters of Carlsbad, Spa, Pyrmont and Seltzer. Prof. Marais made an *Prof. Marais' experiment.* experiment in which he produced the conditions present when soda water is drawn from leaden pipes. He allowed some water holding carbonic acid dissolved under pressure to act on a piece of sheet lead for some time. He afterward tested the water and found that it contained in solution about one grain of carbonate of lead per gallon. It would seem, then, that under ordinary circumstances a pipe ought to last for an indefinite time; but, notwithstanding all that has been said, we do occasionally find pipes that have worn out. Prof. Ripley, in a *Prof. Ripley.* report to the Massachusetts Board of Health, from which I have before quoted, speaks of a specimen which had been in contact with cold water only for a period of fifteen years, which was so corroded in the vicinity of the solder joint as to be eaten through, and along the pipe there was a thick coating consisting almost entirely of the carbonate of lead (with organic matter, a little carbonate and sulphate of lime and a trace of the oxide of iron), which had penetrated the pipe in some places to the depth of one-fifteenth of an inch or more. The protecting carbonate of lime was there, but the protection was not perfect. Here we find another instance of the influence of local causes in defeating the action of general laws. The proof which *Causes which defeat the protective action of the carbonates.* establishes the protective action of the carbonates is ample, and we must look elsewhere to find a reason for occasional excep-

172   THE CHEMISTRY OF PLUMBING.

tions to the general rule. The water is sometimes delivered under great pressure, and other physical agencies tend to impair the strength of the pipe and to promote corrosion.

Sulphates in water. Water may contain several sulphates. I have already spoken of sulphate of soda (Glauber's salt), sulphate of lime (gypsum), sulphate of magnesia (Epsom salt), sulphate of potash and sulphate of alumina. The sulphates, except in a few cases, are not found to such a large extent in potable waters as the carbonates.

Proportion of carbonates and sulphates in the water supplied to five cities. The following table gives the number of grains per gallon of carbonates and sulphates in the drinking water used by the cities of New York, Boston, Philadelphia, London and Liverpool. The fourth column gives by percentage the proportion of the sulphates to the carbonates, showing how much the carbonates exceed the sulphates in their distribution in natural waters:

Name.	Carbonates. Grains per gallon.	Sulphates. Grains per gallon.	Per cent.
Croton................	4·658	0·388	·10
Cochituate (Boston).....	0·830	0·102	·12
Schuylkill (Philadelphia).	3·867	0·057	·01
London..................	10·972	5·765	·53
Liverpool...............	0·870	1·000	1·15

Action of sulphates. Sulphates act like carbonates and protect lead from corrosion. M. Fordos records an interesting experiment which explains the *M Fordos' experiment.* manner in which the sulphates protect lead. He agitated a solution of sulphate of soda in contact with some pellets of lead in presence of air. There was soon formed a white powder consisting of carbonate of lead and sulphate of lead. The following action had gone on in the liquid: The oxide of lead had acted on the sulphate of soda and had abstracted some of its sulphuric acid. This liberated a little soda, which then combined with the carbonic acid of the air and formed carbonate of soda, which in turn was acted on by the lead, forming *Unstable combinations of acids with bases.* carbonate of lead. It must be remembered that such changes as these are constantly occurring, one substance displacing another from its combination and the displaced substance com-

THE CHEMISTRY OF PLUMBING.

bining with something else. It is the course of things which is actually carried out in the water pipe.

In ground abounding in iron pyrites (sulphide of iron), disorganization and oxidation of the sulphide often takes place, the product being sulphate of iron, or copperas. The latter substance frequently finds its way into water, and must be classed as an exception to the rule regulating the sulphates. It indirectly causes a very serious corrosion of the lead. It is true, however, in the case of the sulphates as in the case of the carbonates, that their presence in water does not prevent the water from dissolving a trace of lead. It is merely a scientific fact of no great practical importance, as the quantity of lead dissolved is so small, although a trifle larger than in the case of the carbonates. *Sulphide and sulphate of iron.*

The action of the phosphates, it is agreed, is to prevent the action of water on lead. The phosphate of lead, which is formed when water containing those salts flows through lead pipes, is quite insoluble. The importance of this fact, however, is not very great, as but few waters contain phosphates. The Croton is a remarkable exception, as it contains, according to Prof. Silliman's analysis, 0·832 grains to the gallon, or more than as much as of sulphates. The other waters mentioned in the table contain no phosphates at all except in the case of some of the water supplied to London, which contains a trace of phosphate of lime. The protective influence of the phosphates may be completely destroyed by the presence of organic matter. *Action of phosphates on lead. Large percentage of phosphates in Croton.*

When water flows over iron pyrites it becomes impregnated with a gas very offensive to the sense of smell. It is called sulphur water, and owes its offensive smell to sulphuretted hydrogen. It is claimed that sulphur waters attack lead pipes very vigorously. They form an insoluble sulphide of lead on the pipes; but inasmuch as the sulphuretted hydrogen dissolved in the water is a gas and acts directly on the lead and without the intervention of the formation of the oxide of lead, it is likely that the coating of the sulphide presents no obstacle to the constant corrosion of the lead by the gas. The use of lead pipes for the conveyance of sulphur water is, I think, unsafe. *Sulphuretted hydrogen. Action of sulphur waters on lead.*

174   THE CHEMISTRY OF PLUMBING.

<small>Action of nitrates and nitrites of lead.</small> Waters containing nitrates and nitrites usually attack lead vigorously, forming nitrate and nitrite of lead, both of which are readily soluble in water and very poisonous. They corrode the lead; the resulting salts are washed away, leaving the surface of the lead clean, and the corrosion goes on. Both nitrates and nitrites are formed from the action of organic matter and act in much the same way. For convenience I may call them both nitrates.

<small>Action of nitrates not in proportion to quantity in water.</small> A very small quantity of nitrate of ammonia in water, or of any other nitrate, acts just as vigorously as a large quantity. In some experiments made by Dr. Muir, a grain and a half of a nitrate to the gallon seemed to act as vigorously as double that quantity. If the quantity of nitrates be sufficient, they will corrode a pipe even in the presence of other salts, and in cases in which pipes badly corroded have been examined, a crust of <small>Dr. Muir's experiment.</small> carbonate and sulphate of lead has often been found. Beneath this coating pits in the lead were discovered, sometimes far apart and sometimes close together; sometimes a few in number and sometimes numerous. "The coat over these caverns," says the experimenter, "was generally elevated and mammillary protuberances were thus produced. The action had been most energetic beneath this elevated portion of the coat, the pits being generally bright and of metallic appearance." The action of the nitrates is often facilitated by the presence of certain other salts, such as copperas (sulphate of iron).

<small>Sources of nitrates in water.</small> The source of the nitrates is the decay of animal or vegetable material. This furnishes the nitric acid which combines with the lime, or alkali, or other basis found in the soil, and forms nitrates. Nitrates are largely formed in stables and wherever sewage is allowed to ferment or decay. Wells often contain considerable quantities of nitrates. River water, especially after a freshet, and spring water sometimes contain nitrates, but in fluctuating quantities. The East London Water Company's water contained seven-tenths of a grain per gallon, but this quantity as yet seems to have produced no bad results.

THE CHEMISTRY OF PLUMBING. 175

The chlorides are much less vigorous in their action upon *Chloride* lead than the nitrates and nitrites. Common salt is an example of a chloride. Its chemical name is chloride of sodium. When *Formation of chlorides* soda is dissolved in hydrochloric acid (muriatic), chloride of sodium, or salt, is produced. In like manner, when potash is dissolved in hydrochloric acid, chloride of potassium is produced. This brief explanation may show the meaning of the term chloride, which indicates, without going too deeply into the theory of chemical combination, the union of hydrochloric acid with some base, like potash, soda or lime. Chlorides are abundant in waters, and an exact knowledge of *Chlorides usually in* their action on lead is important. The chlorides found in *excess of sulphates.* water are chloride of potassium, chloride of sodium, chloride of calcium, chloride of magnesium, and, rarely, chloride of aluminum. By reference to a table already given, the reader will see the amount of chlorides per gallon in the water supplied to several large cities. The chlorides are usually present in greater proportion than the sulphates in potable waters. A well at Hartford, Conn., yielded on analysis, $15\frac{1}{4}$ grains of chloride of sodium, $10\frac{1}{2}$ grains of chloride of calcium and $2\frac{1}{4}$ grains of chloride of magnesium per gallon, or 28·398 grains per gallon in all. The Red River contains 38 grains of common salt per gallon. The Hampstead water supplied to London contains as much as 7 grains of salt per gallon. The Trent, of England, holds in solution $17\frac{1}{2}$ grains per gallon. Now, with *Action of chlorides on lead.* reference to the action of these chlorides on lead, concerning which it is very important that we be exact, chemists differ. The generally received opinion has been for a long time that the presence of the chlorides facilitated the corrosion of the lead. The chloride of lead, which is formed by the action of a chloride on that metal, is slightly soluble in water, one part of water taking up $\frac{1}{135}$ of its weight of the salt. This *Solubility of chlorides.* amount of solubility is very dangerous, as the proportion 1 to 135 means $1\frac{1}{4}$ ounces of the poisonous salt to the gallon. On *Muir's experiment.* the other hand, Prof. Muir, after extended research, gave as one of the results of his investigations, in a paper read before the Glasgow Philosophical Society, that the chlorides do not increase, but rather diminish, the action of water on lead, and that, too, when the water contains a nitrate. Perhaps the singular result of the professor's experiments may be explained by considering a fact to which I have already referred, and the disregard of which has been an occasion for stumbling to many experimenters, viz., that some salts which are only partially soluble, or quite insoluble, in water, may have their condition as regards solubility entirely changed by the introduction of another salt. Prof. Muir suspended bright sheets of lead in a solution of chloride of calcium. Had he chosen another chloride his results might have been different. The action of the chloride on the lead produced a chloride of lead which did not dissolve, and consequently the water showed only traces of lead. It is known, however, that the effect of chloride of calcium is to prevent the solution of the chloride of lead already formed; for, although the latter is soluble in 135 parts of water, it takes 634 parts of water containing chloride of calcium to dissolve it. The experiments of Dr. Muir, then, cannot be regarded as disproving the experience of the past.

There is an important difference between the action of the chlorides on lead and the action of the other salts to which I have referred. The latter require that the lead should be first oxidized or rusted before they can act. The chlorides act directly, without the formation of an oxide. No air, therefore, is necessary in such action. It follows from this that the chlorides will act in cases in which other agents fail. The action of the chloride, also, does not stop with the formation of a crust on the lead, as is the case with the carbonates. It is continuous. The process of solution goes on until all the salt is used or the lead entirely dissolved. After a time little white knobs will be found on the lead, varying in size from a pin's head to a pea. If these are removed the lead will be found to be pitted and very bright in these places. But

THE CHEMISTRY OF PLUMBING. 177

besides their direct action, the chlorides have an indirect influence, since they tend to render soluble in water the otherwise insoluble sulphate of lead.

The corrosive action of the chlorides has an especial importance in connection with sea water, or those portions of rivers impregnated with tide water. Sea water usually contains from 2 to 3 per cent. of its weight of common salt. The following table shows the number of parts of the chlorides in 1000 parts of sea water. Two specimens are given, one from the British Channel and one from the Mediterranean: *Chlorides in sea water.*

Salt.	British Channel.	Mediterranean.
Chloride of sodium	28·059	29·424
Chloride of potassium	0·766	0·505
Chloride of magnesium	3·666	3·219
Total chlorides	32·491	33·148

It should be remembered that the corrosive action of the chlorides is not nearly so great as that of the nitrates, for the chloride of lead is much less soluble than the nitrate of lead, and the former may, therefore, be formed on the pipe as a thin coating which, under some circumstances, may act as a slight protection, although it is liable to be, and is, constantly washed away by the dissolving action of the water. *Chlorides less destructive of lead than nitrates.*

The evidence of the action of the chlorides, as deduced from cases of disease, is somewhat uncertain, as in a record of cases of lead poisoning a careful analysis of the water is seldom given. Dr. Christison speaks of a house in Banfshire which was supplied through lead pipe with water from a spring three-quarters of a mile distant. Two and a half years after the owner's occupation of the house began he was seized with severe abdominal complaints, apparently incurable. He left the place and went to Edinburgh, where he recovered. He returned home and began to use the water from the lead pipes, and his disease returned. An analysis of the water showed the presence of lead in it. A more thorough analysis showed the *Evidences of lead poisoning due to chlorides.*

solid contents of the water to be $\frac{1}{16800}$, a large portion of which was chloride of sodium.

Lord Aberdeen's country residence was supplied with water from a spring through lead pipes. Several inmates of the house were presently taken sick with lead disease. A white film was discovered on the chamber water bottle. Treatment for lead colic removed the difficulty. The analysis of the water showed that $\frac{1}{4400}$ of it was solid matter, most of which consisted of chloride of sodium.

Iodides and bromides. The iodides and bromides have an effect upon lead very similar to that exerted by the chlorides, but they are rarely found in potable waters, and then in very minute quantities. They do not need, therefore, to be classed among the agencies of corrosion to which lead pipe is commonly subjected.

Organic matter. The action of organic matter upon lead is usually prompt and positive. By organic matter is meant animal or vegetable sub-
The vegetable acids. stances and their immediate products. This part of the investigation has been to some extent anticipated by what I have said respecting the action of acids; but organic matter does not exist in water simply in the form of vegetable acids.
Sources of organic contamination. In fact, the acids formed by the decay of organized bodies in water are only intermediate compounds. The composition of the organic matter varies with every stage of decay. Changes in temperature, contact of other substances and exposure to the air, induce a constant change in the constitution of such materials. As a consequence, organic bodies of various constitution are found in water.

Occurrence of organic matter in water The organic materials present in water are of two kinds—they occur in a state of solution, or are simply suspended in the water, retaining in a finely divided condition their solid form. They sometimes give the water an acid reaction and sometimes an alkaline reaction.

Soluble organic matter. Soluble organic matter may be derived from vegetable decomposition, in which case there is generally no nitrogen, and, consequently, no ammonia present, or from the decay of animal matter, which gives rise to the formation of nitrogenous compounds (ammonia and nitrates). In fact, the latter class of compounds usually contain, according to Dr. Wm. Proctor, of England, from 2·5 to 7 per cent. of ammonia, a dangerous constituent in respect to the action of the water on lead. River water usually holds in suspension or solution a considerable quantity of matter of animal and vegetable origin. Such are weeds, fish spawn, leaves, mud and microscopic animals. The decomposition of these bodies produces organic compounds. Animal and vegetable matter in river water.

The action of such substances is most important. If they are simply held in suspension, they may generally be kept out of the pipes by a proper method of filtering. If they are allowed to enter the pipes, they are apt to lodge in some bend or angle, where they form a nucleus around which other organic matters may collect. Here they decompose and form compounds which dissolve away the protecting crust of carbonate of lead and corrode the pipes. By their decomposition they evolve ammoniacal compounds and nitrates, both of which are destructive to lead. I have already referred to the action of the vegetable acids and alkalies. If the organic matter is dissolved in the water instead of being held in suspension, it cannot be kept out by means of filters, and if it is present in large quantity, it renders the water dangerous to the safety of lead. Action of organic matter on lead.

The process of decay, or fermentation, gives rise to the element of danger in the presence of organic matter in the pipes. Decay, or fermentation, is simply the decomposition of a body into its constituents, and the recombining of these constituents in new forms with new properties. It is a chemical principle generally conceded that an element, in passing from one state of combination into another, is most active in its properties. So generally recognized is this principle that the adjective *nascent* (meaning new-born) is applied to a body in such a condition. In the process of decay many substances Decay and fermentation of organic matter in pipes.

must be passing into their nascent condition, being liberated by decomposition, and out of it again when a new combination is formed by the free substance. During this nascent condition the free substance is likely to attack anything which may be present and for which it has an affinity. Whenever, therefore, lead is exposed to contact with fermenting matter, it is rapidly oxidized, and the oxide thus formed is dissolved by the organic acids which result from the fermentation. Even if the salt formed by the organic acid and the lead is insoluble in water, it may be dissolved by an excess of the acid in the water. The principle in accordance with which this takes place has already been referred to. Lead is often corroded by contact with decaying wood. In Amsterdam lead roofs were substituted for tiles. The inhabitants used, for culinary purposes, water collected from the roofs and through lead gutters. Lead colic, which had rarely appeared in that city, broke out as soon as the lead roofs were introduced, and in a violent form. Doubtless the purity of the rain water had much to do with the result, but as the trouble occurred especially in the autumn, the inhabitants ascribed the rapid corrosion, in great measure, to the decaying leaves which at that season lie on the roofs.

Corrosion of lead by contact with decaying wood.

Lead poisoning in Amsterdam.

An illustration of the action of organic matter on lead may be obtained from a little experiment which any one may perform. If a strip of bright lead be immersed in a glass of dark-colored rose water exposed to the air, the water after some time will usually be rendered colorless. The organic matter of the rose water is decomposed by the lead, which itself during the process is corroded. If such water be tested afterward it will be found to contain lead in solution.

Experimental illustration of the action of organic matter on lead.

One of the most important features of the action of organic matter on lead is the fact that it enables water to dissolve some of the protective salts of lead, such as the sulphate and phosphate.

The amount of organic matter in water is variable. There is generally a small amount in all waters. Even water from

Amount of organic matter in water.

granitic districts, according to Dr. Proctor, contains from 0·3 to 0·7 grains per gallon, while water which has permeated vegetable soil may afford 12 to 30, or even more, grains per gallon. The amount of organic matter generally depends on local causes. The sources of organic matter derived from animal decay are numerous, chiefly animal excreta and the refuse of manufactories. The contents of sewers and cesspools drain into springs or rivers, or else the water permeates the soil more or less impregnated with sewage. Water may be contaminated in this manner by a nuisance at a considerable distance from it, depending on the porosity and tenacity of the soil. There is very little doubt that to this cause many cases of corrosion of pipes and many accidents are due in localities where the constitution of the water and the general experience point to safety in the use of lead pipes. Neither the Croton nor Cochituate water usually contains organic matter. The Schuylkill water, on an analysis by Prof. Silliman, Jr., showed 0·08 grain to the gallon. *Sewage contamination.*

We have considered the action of each salt as though a water could be found which contained only one salt in solution. The fact is that waters often contain several salts, and the question naturally arises, May not one salt interfere with or affect the action of another? On this question Prof. Muir advances some opinions based upon experimental tests. He poured into a clean flask 500 c. c. (about one-tenth of a gallon) of water, and poured a similar quantity into each of several other flasks. " To these were added weighed quantities of various salts. Pieces of clean, bright lead were then suspended by threads in these solutions, so that the liquid should have free access to all parts of the lead. Thus the surface of lead acted on could be accurately determined. Each piece was of the same size, and the surface acted on was 8·65 square inches. The flasks were set aside for 24, 48 and 72 hours, and at the expiration of each period the amount of lead dissolved was estimated." An idea of the accuracy of the operation by which these amounts of lead were *Action on lead of water containing mixed salts.* *Muir's experiments with mixed salts.*

THE CHEMISTRY OF PLUMBING.

estimated may be gained from the fact that the reaction employed was sufficiently delicate to detect two parts of lead in 1,000,000 parts of water. He gives the results of his experiments in tabulated form, and from this table I take so much as refers to mixtures of salts, changing the arrangement for the sake of greater clearness. The first column gives the number of the experiment; the second gives the names of the salts placed in the flask in which the experiment was carried on; the third gives the proportion of the salts (the strength of the solution) in grains per gallon; the fourth, fifth and sixth give the estimated amount of lead dissolved in the water at the end of the three periods, 24, 48 and 72 hours.

Results.

Number of Experiments.	Names of Salts.	Grains of salts per gallon.	Grains of lead per gallon dissolved in 24 hours.	48 hours.	72 hours.
1	Nitrate of ammonia................	1.4	0.91	....	1.75
2	Nitrate of potash................ Sulphate of soda................	1.4 3.5	0.14	0.14	....
3	Nitrate of ammonia................	2.8	1.05	1.05	2.24
4	Nitrate of potash................ Sulphate of soda................	2.8 14.8	0.05	0.07	0.08
5	Nitrate of potash................ Carbonate of potash................	3.1 21.7	....	....	0.021
6	Nitrate of potash................ Sulphate of potash................	5.4 35.2	....	....	0.035
7	Nitrate of ammonia .... Carbonate of potash.... .... Sulphate of soda................	1.4 7.0 14.0	...	....	0.028
8	Sulphate of soda................ Carbonate of potash................ Chloride of calcium................	14.0 2.8 7.0	....	....	0.007

Examination of Muir's table. The table, although divested of much of its original intricacy, seems somewhat complicated, and requires a little study in

THE CHEMISTRY OF PLUMBING. 183

order that we may see the important practical truths which it reveals. In experiment No. 1 a nitrate alone was used and its corrosiveness noted. The quantity of nitrate used, it must be remembered, is very large; we seldom see any natural water with so large a proportion. The East London Water Company's water, previously referred to, only contained one-half this quantity, or 0·7 grain per gallon. In experiment No. 2 the same quantity of nitrate was used and a sulphate added, the proportion being as 1 to $2\frac{1}{2}$. Mark the protecting power of the sulphate. After 24 hours the nitrate alone had removed 0·91 grain, but when accompanied by the sulphate it only removed 0·14 grain. In experiment No. 4 the proportion of nitrate to sulphate was 1 to 6. In this case only 0·05 grain was removed after 24 hours and only 0·08 grain after 72 hours. In experiment No. 3, where the same quantity of nitrate was used as in No. 4, 2·24 grains were removed at the end of 72 hours. Experiment No. 5 shows that a carbonate exercises a more powerful protective influence than the sulphate, for after 48 hours the nitrate had removed no lead, and only 0·021 grain after 72 hours, practically nothing. In experiment No. 7, in which a nitrate, carbonate and sulphate were used, there was no action until after the third period, and then only 0·028 grain had been removed. In experiment No. 8 a chloride was substituted for a nitrate.

The results of these experiments are very important, since they teach that even the dangerous nitrates may exist in water without any detriment to the pipe if there be also sufficient carbonates and sulphates. If the proportion of nitrate to carbonate or sulphate is large, the latter salts offer but little protection to the pipe, nor can we expect any favorable result if the chlorides be in excess. Sulphate of lead is somewhat soluble in water containing chlorides. A case is reported by Dr. Thomason of the poisoning of a number of people at Tunbridge, England. The water was conducted a quarter of a mile through a lead pipe. An analysis showed that it was very pure, *Importance of Muir's experiments.* *Lead poisoning at Tunbridge.*

containing only one part of saline matter (three-fourths chloride of sodium) in 38,000 parts of water. In this case the proportion of the chloride was too great. The action of mixed salts would seem, therefore, to depend upon their proportion to each other in the water holding them in solution. When the proportion of the corrosive salts is not too great they may be assumed not to interfere with the action of the protective salts.

Action of mixed salts dependent upon their mature proportions.

The water supplied to the city of London is a practical illustration of the truth of these conclusions. It contains both nitrates and chlorides, and yet, on account of the abundance of carbonates and sulphates, the influence on the lead pipes through which it flows is unimportant.

London water supply.

When organic matter is present in the water, in connection with other substances, it is almost impossible to predict what will be the action. It is safe to say, however, that unless the quantity of organic matter is exceedingly small, the pipes which convey the water should be suspected until a long experience has proven that there is no danger. The general experience is unfavorable, the organic matter in many instances completely destroying the protective action of the other salts. The crusts of carbonate, sulphate and phosphate of lead, which owe their protective action entirely to their insolubility, are often easily dissolved by an excess of organic matter in the water. Compounds of ammonia with organic acids easily dissolve sulphate of lead. Phosphate of lead is readily dissolved by the feeblest acids.

Action on lead of organic matter in combination with other substances.

The most vigorous action occurs when several corrosive salts are found together in water, or are found in connection with organic matter. The most usual combination is that of the chlorides and organic matter. Chloride of lead is soluble to a dangerous extent in water, as we have seen, but its solubility is much increased by the presence of vegetable acids, and of ammonia and other alkalies.

Mixtures of corrosive salts.

Sulphate of iron (copperas) and nitrate of lime (lime saltpeter) react on each other in a manner dangerous to lead pipe. Copperas is composed of sulphuric acid and iron. When

Examples of corrosive salts in combination.

THE CHEMISTRY OF PLUMBING. 185

exposed to air and water the iron becomes rusted, and this reaction sets free some of the sulphuric acid, which then acts on the nitrate of lime. The latter substance is composed of nitric acid and lime. The free sulphuric acid attacks the lime and expels the nitric acid, which, being set free, attacks the lead.

Extraneous substances not belonging to the water which flows through the pipe sometimes act as corroding agents, and their action has often been confounded with the action of the water itself. I have already alluded to the effect of a piece of mortar dropping into a cistern or tank, and also to the possibility of outside corrosion of a pipe. I shall now refer to only one other extraneous substance which may affect the integrity of lead conduits. Service pipes of this metal are generally attached to iron mains, and iron rust is sometimes carried from the main to the pipe. The question is, Does this substance affect the pipe? Authorities are diametrically opposed in their answers to this question. In the first place, we have an experiment made by Prof. Horsford. Iron combines with the oxygen of the air in several proportions. Iron rust is an oxide of iron which contains one and a half times as much oxygen as the oxide of iron which usually enters into combination of acids to form sulphates, phosphates, &c. The latter oxide is called the protoxide. The theory of the action between lead and iron rust is that lead takes away a portion of the oxygen from the iron rust, reducing it to protoxide of iron, which combines with other materials in the water. When the lead withdraws the oxygen from the iron and appropriates it to itself, it becomes oxidized or rusted. This theory of the oxidation or corrosion of the lead Prof. Horsford disputes, and defends his position by experiment. He placed bars of lead in contact with iron rust, in open tubes containing Cochituate, Croton, Jamaica, Fairmount, Albany and Troy water, and at the end of two days tested the water for protoxide of iron. No reaction was obtained. Subsequent tests were made at the end of seven,

[margin: Influence of extraneous substances in water.]
[margin: Union of lead and iron pipes.]
[margin: Prof. Horsford's experiments.]

twelve and twenty-three days, but no trace of protoxide of iron was found in the water. The iron rust had, therefore, not been reduced by the lead. He again placed iron rust and bright bars of lead in flasks of distilled and Cochituate water and sealed them. These flasks were kept for a long time and the brightness of the lead was not in the slightest degree dimmed. As a laboratory experiment this is of importance, but in the light of the actual experience of lead pipe in contact with iron, it seems insufficient. The following case, although not so systematic nor so easily explained, is much more important. The water of a certain spring in England had flowed into and from a leaden reservoir for 60 years without injury to the reservoir or contamination of the water. It was conveyed to and away from the reservoir in lead pipes. The water was afterward conveyed through iron pipes and immediately lead was found in solution. The water was then found also to be so destructive to the bottoms of lead cisterns that some of them had to be renewed in five or six years.

Examples of lead corrosion by iron rust.

It would seem that an extreme inference in either direction regarding the action of iron rust on lead would be inadvisable. On the one hand, it must be remembered that it is impossible in a laboratory experiment to reproduce all the conditions which exist in the lead pipe. For instance, the rust in the pipe is carried along with rapidity by the water flowing through the conduit, and physical action may have something to do in facilitating the corrosion of the lead by the iron. Again, the question of the influence of other substances in promoting chemical action must not be forgotten. In the experiment of Prof. Horsford, there were used simply iron rust, lead and water containing other substances in solution. Suppose, however, that different waters had been used, holding in solution different substances, or that another quality of lead had been employed, other results might have been expected. It is advisable to suspect iron rust until by long experience its harmless character in each particular case is established. Its action no

Sources of error in laboratory experiments.

doubt varies with the constitution of the water and the circumstances of physical contact which attend its presence in lead pipe.

We leave now the subject of the corrosion of the lead, as affected by the peculiar constitution of the water, to consider certain influences which either act independently of the substances that are dissolved in the water or, when an erosion of the lead has already commenced, tend to increase and hasten the process. These influences have perplexed experimenters more than other branches of the subject. They bring about anomalous and unexpected results, often in direct antagonism to some pet theory which has been built up on the subject. To this may be attributed many of the contradictions in the results of the observations of various chemists. *Corrosion of lead by influences independent of the constituents of water.*

Prominent among these disturbing influences is galvanic action. No elaborate apparatus is necessary to bring this agency into operation. The lead pipe with its metallic connections is often a battery in itself. All batteries depend for their action upon the contact of two metals immersed in a bath of some liquid. Usually, the two metals employed are zinc and platinum or silver, and the bath in which they are immersed is dilute sulphuric acid. The acid attacks only one of the metals, zinc, and leaves the other intact. The metal attacked by the acid is called the positive metal, and the other the negative metal. Whenever two metals are brought together a galvanic action ensues. The intensity of the effect varies with the electrical character of the metals used. Certain metals, as tin and copper, have but a weak action, while others, as silver and zinc, produce a vigorous galvanic current. The truth of the latter statement can be ascertained by placing a silver coin on one side of the tongue, and a strip of zinc on the other, and bringing the exteroir edges together. A sharp, prickly sensation is felt. Here we have the two metals, while the fluid is saliva. Let it be remembered that the contact of two metals is the simple principle which, elaborated and applied to convenient ap- *Galvanic action. Conditions of galvanic action.*

paratus, underlies the whole subject of galvanic electricity, and it will be easy to make the application to lead pipes.

Positive and negative metals. When a current of electricity is thus excited by the contact of two metals in saline solution, it has been found that only the positive metal is usually very much corroded, while the other, the negative metal, is unharmed. Lead pipes are often placed in contact with iron, copper or tin. In every such case we have a galvanic battery, the two metals in contact being the poles, and the water flowing through the pipes the saline solution. As a consequence, a galvanic current is excited and corrosion takes place. The nature of the salts dissolved modifies this action. Dr. Christison says: "The presence of bars of other metals crossing lead, or bits of them lying upon it, will also develop the same action." It is possible, also, that iron rust, and even the carbonate of lead compounds which encrust the pipe, may sometimes be thrown into electrical relations with the pipe.

Christison on galvanic action in lead pipes.

These statements do not present a new fact. The corrosion of lead by galvanic action, caused by uniting lead with other metals under water, was proven long ago by the experiments of Dr. Paris, England. The importance of this fact may be illustrated by reference to what has probably occurred frequently in the case of iron ships, namely, the corrosion of the iron plates by galvanic action developed by the contact of copper and iron.

Corrosion of ship plates by galvanic action.

The Engineer, a few years ago, in commenting on the wreck of the ironclad Megara, called attention to the startling fact that, should even a minute piece of copper remain in contact with the inside bottom plates of an iron ship, in a bath of bilge water, as under the circumstances of the case it necessarily must be, an active galvanic energy is established between the two metals, and the iron being the sacrificial metal (*i. e.* the positive metal), " the bottom will sooner or later be eaten through with a hole somewhat larger than the superimposed copper."

The Megara.

Electrical relations of metals. The relation of the metals to each other in producing currents of electricity varies somewhat with the constitution of the water. The order is as follows:

THE CHEMISTRY OF PLUMBING. 189

In Acid Waters.	*In Alkaline Waters.*
Zinc,	Zinc,
Tin,	Tin,
Lead,	Iron,
Copper,	Lead,
Iron,	Copper.

In the above columns each metal is positive to all below it; that is, if any two of the metals are in contact with each other in a solution, the one that stands above the other (in the table) will be consumed. For example, were iron and lead brought in contact in an alkaline solution, the former would be rusted at the point where the latter touched it.

Copper and brass often come in contact with lead where copper couplings, boilers or faucets are used, or, in some cases, where copper screws, bars or pipes touch the lead. That galvanic action often occurs under such circumstances is unquestionable, and as to which of the two metals is corroded there is but little doubt. By consulting the tables above, we find that lead, being above copper, is the positive metal, and must, therefore, be corroded. Some interesting experiments were made by Mr. F. Casamajor, throwing a very clear light on this point. He took four glass flasks, into each of which he poured about one-fifth of a pint of aqueduct water drawn fresh from the hydrant. In two of these flasks he placed pieces of sheet lead, perfectly clean, the surface of lead in each flask being three square inches. In the other two flasks he placed little bundles of sheet lead and copper wire rolled up together in perfect contact. The lead and copper had each a surface of three square inches, and each was perfectly clean and bright. One flask with lead alone, and one with lead and copper, were left in a dark place for forty hours at a temperature of 75° F. The other two flasks, one with lead alone, and one with lead and copper, were left in a dark place for 40 hours at a temperature of 150° F. The object of placing the flasks in the dark was to approach the actual condition of things in lead pipe. At the expiration of

[marginal notes: Copper and brass in contact with lead. Casamajor's experiments.]

40 hours both flasks containing only lead and water were examined. The lead was perfectly bright and the water limpid. On testing the water no lead was discovered. The two flasks containing the lead and copper presented a very different appearance. The surface of the lead was coated with a white oxycarbonate, which, on shaking the flask, spread through the water, making it turbid. The water was tested for copper and showed the faintest trace. Another portion of water from the same flask gave the usual reactions for lead.

Example of lead corrosion by galvanic action. So much for the laboratory. A practical illustration will confirm the truth of the inference to be drawn from the above experiment. A gentleman residing near Baltimore, having occasion to have a pump repaired, on examining the leak found that the lead pipe, which was connected to the pump by a brass coupling, was almost destroyed in the vicinity of the brass. The corrosion extended for an inch from the coupling, and the pipe was held together by a few shreds of lead. The pipe, which was used to carry water from a well, was entirely uninjured in every other portion. Dr. R. Buckler, who reports this fact, made an experiment upon the water of this well to satisfy himself of the cause of such a remarkable corrosion. He placed about four ounces of the water in two beaker glasses, and in one immersed a bright strip of lead; in the other he immersed a strip of lead and brass connected. The beakers were covered with paper, so as not to exclude the air entirely, and allowed to remain undisturbed for a week. He then tested water from both glasses. In the case of the water containing only lead he obtained a slight precipitate, and in the other a copious precipitate. The fact that where copper or brass and lead are in contact under water a galvanic action is liable to occur, promoting corrosion, is, therefore, pretty clearly established.

Dr. Buckler's experiment.

Contact of tin and lead. Tin is very frequently used in contact with lead. Solder is an alloy of lead and tin, and it is well known that corrosion is generally the greatest in the vicinity of the solder. The cause is galvanic action, which is likely to occur when the two metals

Joints in tin-lined lead pipes.

composing lead-encased tin pipe are both subjected to the action of water; hence the importance of making such joints as will insure a continuous tin lining where this pipe is used, and the danger attending the practice of merely bringing the ends together and wiping the joint with solder.

A great variety of physical influences operate either to cause or accelerate the corrosion of lead by water. These influences often act in direct antagonism to the agencies which protect the pipe, and in the case of anomalous action of a certain water, the explanation is often to be looked for in this direction. They come under the head of local causes. *Action of physical influences on lead.*

A defect in the pipe will sometimes promote corrosion. Dr. Dana remarks that he has seen lead pipe which conveyed spring water very much eroded and slightly perforated by the enlargement of an original defect in a part of the pipe. The water was very pure, and yet, "notwithstanding a deposited coat of oxide and carbonate of lead, the erosion continued and lead was dissolved even at the end of eight years." *Defects in pipes.*

Strains on the pipes will produce unfavorable results. Water is often delivered under great pressure, and seams are by this means made in the pipes, and a disarrangement of the particles ensues. Where the circumstances are favorable, the corrosion is promoted by a combination of mechanical and chemical forces. The freezing of water, when it does not burst the pipe, has the effect of straining it, thus producing a change in the molecular arrangement of the material of the pipe and facilitating corrosion. A strain on the lead produced by bending it sharply, will sometimes facilitate corrosion, and for the same reasons as in the instances just named. Prof. Nichols records a case in which this cause is clearly indicated. He says that he had in his possession a section of supply pipe "removed from the aqueduct of a neighboring city (*i. e.*, near Boston), in a portion of which corrosive action had proceeded so far as to cause leakage. The part thus acted upon was confined to an acute angle, and there is evidence that the plumber, in placing it in its posi- *Strains on pipes.* *Effects of sharp bends.*

tion, bent it in the wrong direction, thus creating the necessity for another turn. This pipe had, doubtless, been subjected to two violent turns, which seriously impaired the homogeneity of the metal." Prof. Nichols suggests that the disturbance of the crystalline structure of the metal by strains may change its electrical condition, and that thus galvanic action may be promoted, giving rise to chemical decomposition.

Prof. Nichols' theory.

A few simple facts should be here noted, and, indeed, the omission to give them proper weight may be regarded as one of the causes of the numerous complexities and contradictions with which chemists have surrounded the subject of the action of water on lead. First, there is a great deal of difference in the activity of corrosion in new and old pipes. No matter what the tendency of a water to form an insoluble protective coating on leaden surfaces, there is usually some corrosive action at first. Croton water is commonly regarded as exerting no dangerous corrosive action on lead, but when Croton pipes are taken up at the end of several months they will be found to be corroded, and in some places the lead is pitted. Dr. Dana remarks that he has examined small sections of pipes which had been used in conveying water from the James River, Va., for twelve years. There was a "fine, reddish colored and quite smooth and compact coat deposited on the inner surface of the pipe, which was easily detached, showing evident and unmistakable marks of corrosion by small pits and thread-like channels." These were evidently made when the pipes were new and before the crust formed.

Activity of corrosion in new and old pipes.

It is interesting to enquire how long a time must elapse before a lead pipe becomes sufficiently incrusted to admit of its being used with safety. Prof. Nichols immersed a section of a new lead pipe in Cochituate water for one hour at the temperature of 65° F. The water then gave decided evidence of lead being present. The piece was removed and placed in six fresh portions of water, one hour in each. Each sample of water gave the reaction of lead. The experiment was continued for

Duration of corrosive action in new pipes.

THE CHEMISTRY OF PLUMBING.

two weeks, varying the time of immersion in fresh portions of water from one hour to ten. The lead indications still continued, although at last they were feeble.

The amount of lead which water exercising a corrosive action is found to contain, is usually dependent upon the length of time it has stood in the pipe. It is also affected by temperature and season. Waters from given sources do not show the same constitution from one season to another. An excess of rain may temporarily change the character of the water and its action on lead; while an elevation of its temperature may make a water previously harmless extremely dangerous as regards its action on lead. *Amount of lead which water will contain.*

From the facts already presented in this chapter, the reader will, in all probability, be led to the conclusion that lead pipe can never be used without giving rise to the danger of lead poisoning. I think it safe to venture the opinion that the danger exists in some degree under all but the most exceptional circumstances. Admitting that severe cases of lead poisoning rarely occur from the use of water drawn through lead pipes, they do occur, and under a great variety of circumstances; and I have no doubt that nine-tenths of the mischief done by lead in its effect on the human system escapes the notice of physicians. It is undoubtedly true, however, that certain waters can pass through lead pipes without practical contamination, even though they take up enough of the metal to give a lead reaction when subject to the delicate tests of the laboratory. I have already spoken of the manner in which water first acts upon lead and the effects of long-continued contact between them; but it is desirable to recur to this, for the reason that it seems to be the central fact of the whole subject and should be thoroughly understood. The first result of the contact between water and lead is the formation on the surface of the metal of a whitish crust or scum of oxide of lead, formed by the combination of the oxygen of the air dissolved in the water with the lead of the pipe. The next result is the solution of this oxide *Conclusions respecting lead pipes.* *Lead poisoning by water.* *Summary of the facts respecting the action of potable water on lead.*

13

of lead by the water and its removal, if nothing prevents. If this were all, the destruction of lead pipes and the poisoning of those who use them would only be a matter of time, for as fast as the soluble oxide was dissolved away a new coating would be formed. But as soon as the scum of oxide of lead is formed on the surface of the lead, the former is attacked by the carbonic acid (which is always present in the air and is almost always in solution in the water) and the lead is converted into carbonate of lead, one of the principal ingredients of the painter's white lead. This carbonate of lead is practically insoluble, and adheres to the surface of the lead as a hard crust which soon thickens until it prevents the action of the water altogether.

Expedients for guarding against lead poisoning. Fortunately for those who are compelled to use water supplied through lead pipes, chemistry is not without resources for guarding against lead poisoning. The first and simplest of these is filtration. Filters made of chalk have been strongly recommended, and the experiments of M. Robierre are sufficiently minute to warrant us in accepting with confidence his conclusions concerning the efficacy of this material. He states very positively that his researches have led him to the conclusion that the greater portion of poisonous lead compounds in water, obtained by the contact of the common water with lead pipes, is in suspension, and that frequently the filtration of this water through chalk deprives it of its poisonous properties.

Robierre's experiments with chalk.

Wood charcoal filters. Wood charcoal, coarsely pulverized, has also been recommended. This will remove as much as 7 grains of lead to the gallon.

Protection of pipes against corrosion. It is better, of course, to prevent the water from acting on the pipes at all, and to secure this desirable immunity from corrosive action various plans have been suggested. Among these may be mentioned the filtration of water before it enters the pipes. Sand, clay and, better still, animal charcoal answer admirably well as filters, for they not only remove the mechanically suspended particles, but a portion of the organic matter dissolved in the water. The latter result is probably effected by the oxidation of the organic substances into harmless com-

Filtration of water before passing through lead pipes.

Beneficial results of filtration.

pounds while passing through the filters. Soluble organic matter may be almost entirely removed by filtering the water through black or magnetic oxide of iron. This occurs throughout the United States as an ore, and when coarsely ground is an admirable filtering material for organic matter. One of the most remarkable characteristics of this oxide of iron as a filtering material is that it does not perceptibly lose its power by time and use. In the water works of Southport, England, a filter bed was in use for seven years without showing any diminution of power, and in domestic filters used for the same length of time there seems to have been no occasion for cleansing them. *Magnetic oxide of iron.*

When filters cannot be depended upon, the best method of preventing the contamination of water with lead salts is to line the pipes with an insoluble coating. There are various plans for accomplishing this result. One of the oldest, as well as the simplest, is that recommended by Dr. Christison. He says that "a remedy may be found in unusually pure spring water by leaving the pipes full of the water for a few months without drawing off." The water acts on the lead, forming the insoluble coating of carbonate of lead to which I have already referred so frequently. During the long period in which the water is standing in the pipes the coating becomes hard and thick enough to resist the further corrosion of the water. Where this method will not answer, some material should be put in the water in the pipes which will form an insoluble coating on the lead. Dr. Christison recommends phosphate of soda. In one of his experiments he put in some lead pipes some phosphate of soda, in weight about 1·25000 the weight of the water. This would require the use in a pipe 100 feet long and three-fourths inch inside diameter of only 6 grains of phosphate of soda. Fourteen days afterward the solution was discharged and spring water readmitted, and a great improvement had taken place. The solution was replaced and another trial was made six weeks afterward, and the lead could scarcely be *Insoluble linings for pipes. Dr. Christison's method. Phosphate of soda.*

discovered in the water. There was no further trouble with regard to lead in the water, although for more than a year previous the water had continued to act vigorously on the lead.

Rolfe & Gillet's method. Another plan was suggested by Messrs. Rolfe & Gillet, of Boston, which is very effective and works more rapidly than the preceding. They dissolve one pound of sulphide of potassium in two gallons of water, and allow a solution of this strength to remain in the pipes for twelve hours. The interior of the pipes becomes covered with a black impervious coating of sulphide of lead, which prevents the further action of the *Dr. Schwaz's method.* water. Dr. Schwaz, of Breslau, advises the use of a warm and concentrated solution of sulphide of potassium. This will do the work more rapidly, sometimes in fifteen minutes. The more concentrated the solution, however, the more expensive is the operation. Crude sulphide of potassium and sulphide of sodium are sometimes used.

Mr. Perry's method. Another plan was suggested, about five years ago, by Mr. Robert P. Perry, of Newport, R. I. He employs a solution of chromate of potassa, which is poured into the pipes to be protected. It forms an insoluble coating of chromate of lead, which, the inventor claims, protects the pipe and does not interfere with soldering.

Detection of lead in water by analysis. There are several methods of determining by analysis whether lead is present in water or not, all of which are within the ability of the plumber of good general intelligence and judgment who will provide himself with the necessary chemicals and apparatus. Chemical knowledge is not requisite, but neatness, careful manipulation and accuracy in noting results *Delicacy of the reactions,* reached are indispensable. It should be remembered, however, that the reactions by which the presence of lead is determined are very delicate, and any carelessness may cause one to blunder and arrive at false conclusions. It should also be remembered that although the positive results of chemical *Negative results inconclusive.* analysis are conclusive when they show the presence of lead in water, negative results do not always prove that no lead is pres-

THE CHEMISTRY OF PLUMBING. 197

ent. For this reason I should advise the employment of a **Limitation of amateur analyses.** chemist to make careful analyses whenever lead poisoning is feared or suspected. The intelligent plumber or any one else may tell, however, whether water has taken up lead during its passage through pipes or when held in tanks lined with that metal, and also — approximately, at least — whether a given water is likely to act on lead, and the information thus gained may often be of great value. In the succeeding chapter some simple rules are given for determining the constitution of water. We will here direct our attention merely to tests for metallic salts taken up by water in its passage through metallic pipes and reservoirs.

The first step in the analysis of water is to concentrate it. **Concentration of water for analysis.** Draw from the pipes about five gallons of water. The best time to do this is after the water has been standing in the pipes for two or three hours, as there is likely to be more lead in the water than when the water is constantly running. This quantity of water should be boiled until it is evaporated down to a gill or less. In boiling and in all the other operations the experimenter must be careful to use no vessel containing lead. The first portion of the operation may be performed in a large **Method of concentration** vessel and over a stove. Care should be taken to cover the vessel so that no impurities get into the liquid from the air. When the water is reduced in bulk to less than a quart, it should be transferred to a quart glass beaker. Such a vessel can be purchased at any glassware establishment. The beaker, being a little over half full of water, is placed upon a sand bath while the latter is cold, and a pinch of acetate of ammonia is put into the water. An alcohol lamp or gas flame is then placed under the sand bath. The latter is simply a sheet-iron saucer full of sand. The sand distributes the heat along the bottom of the beaker. The object of this concentration is to strengthen the solution. Whenever lead is found in water it is usually in such small quantities that the reactions are very faint, but by boiling the water down we obtain in a gill of water all the lead dissolved in five gallons, and thus have a solution in which we are able to discover lead if any is present.

Filtration. If, after this concentration, the liquid appears turbid, it is best to filter it. Cut filter papers can be bought in packages from dealers in chemical supplies. They consist of circular pieces of a material resembling blotting paper. These are folded and placed in the funnel, which must be of glass; the liquid is poured into the paper cone held in the funnel, and trickles from it clear and limpid. No more water should be poured into the funnel than the cone of filtering paper will hold, and it must, therefore, be supplied in small quantities until the whole of the water is filtered.

We are now ready to test for lead. This may be performed by any of the following methods:

Sulphuretted hydrogen. *Sulphuretted hydrogen.*—This test is so delicate that one part of lead can be detected in 500,000 parts of water. If sulphuretted hydrogen be added to water containing one-tenth grain per gallon, a brownish color is produced. If the water has been concentrated by evaporation to 1-100 of its original bulk before adding this reagent, the thousandth part of a grain in a gallon can, with a little practice, be detected. During the evaporation acetic acid must be added to dissolve the oxycarbonate formed. A small quantity of a solution of citrate of ammonia or of acetate of ammonia is added to dissolve any sulphate of lead that may have been formed. It is very difficult to obtain acetate of ammonia in a solid state, as it requires to be crystallized under the receiver of an air pump, so deliquescent are the crystals.

Sulphuretted hydrogen gas is obtained by the action of dilute acids upon sulphide of iron, sulphide of antimony, or sulphide of potassium (*hepar sulphuris*). It can also be obtained in an impure state by heating together paraffine and sulphur. The gas should be washed by passing through water, and may then be passed directly into the liquid to be tested, or dissolved in water and bottled for subsequent use.

THE CHEMISTRY OF PLUMBING.

Sulphide of potassium, or liver of sulphur, can be employed as a reagent for detecting lead. Its solution produces a dark color in water containing two-thirds grain to the gallon, provided very little of the reagent is added; if more is added, sulphur is precipitated and conceals the lead reaction. *Sulphide of potassium test.*

Sulphide of ammonium (yellow) produces a change of color, perceptible by comparison when only one-third grain of lead is present in a gallon of water. Both this reagent and the one last mentioned also produce black precipitates in water containing iron, but not in water where tin alone is present. Sulphuretted hydrogen, on the contrary, gives a black precipitate with tin, but not with iron. All three of these reagents possess a vile odor and do not keep well. As soon as the odor becomes faint they are useless. *Sulphide of ammonium test.*

Bichromate of potassa.—This salt possesses several advantages over those previously mentioned. It has no odor, can be kept for years either in solution or in crystals, is easily obtained in any drug store under the name of potassiæ bichromas. The saturated solution has a deep red color, but when added to a strong solution of lead a beautiful precipitate of chrome yellow is formed. This precipitate, when treated with nitric acid, turns to a bright red, "chrome red." The addition of bichromate of potassa to water containing one-tenth grain to the gallon produces a change of color easily detected by comparison. In this, as in the former cases, the test should be made as follows: Two test tubes of equal caliber are taken in the left hand; a few drams of pure water are placed in one and an equal volume of the water to be tested in the other. A few drops of the reagent are added to both, and the tubes held in various positions against white and dark backgrounds, against the light and in the shade, viewed vertically and horizontally, until we are convinced that no change has taken place; then a little more of the reagent is added, and so on. These precautions are especially necessary where colored reagents are employed. My own experiments convince me that bichromate of potassa is *Bichromate of potassa test.*

quite as delicate a test for lead as sulphuretted hydrogen when these precautions are observed.

Sulphuric acid test. *Sulphuric acid* and solutions of the sulphates produce a white precipitate with lead, and, according to Lassaigne, one part of lead in 25,000 of water can be detected in 15 minutes by the use of sulphate of soda. As lime also gives a white precipitate with sulphuric acid, this test is not applicable to water in general.

Iodide of potassium test. *Iodide of potassium* produces a yellow precipitate in lead solutions if not too dilute.

Action of carbonic acid on lead in water. When water containing lead is exposed to the air, the carbonic acid of the atmosphere converts the lead into the hydrated oxycarbonate, which is the most insoluble of all the lead salts— so much so that only one part will dissolve in four million parts of water, or one-sixtieth grain per gallon, and hence water which has been exposed to the air a few hours will not contain over $\frac{1}{4000000}$ of lead in solution. If, however, the water contains free carbonic acid, this salt will be dissolved by it, but is *Advantages of boiling water containing lead.* precipitated by boiling. From this it will be seen that persons compelled to use water containing lead may reduce the danger to a minimum by boiling, allowing to stand exposed, and then filtering, or even decanting.

Lead poisoning. In concluding these somewhat extended remarks on lead corrosion, a few words on the subject of lead poisoning may not be without interest for the general reader, and especially for the *Lead pipes not the only cause of lead poisoning.* plumber. I am not disposed to underestimate the danger of conveying water through lead pipes; but it is only candid to admit that lead pipes have probably had to bear the blame of many cases of lead poisoning with which they have had nothing to do. In a locality in which lead is used as a material for conveying water, whenever a case of this disease arises suspicion is generally directed against the water pipes. There are, however, many methods by which lead can be and is unconsciously intro*Snuff.* duced into the system other than in drinking water. Dr. Hassall explained some time ago, in the London *Lancet*, that par-

THE CHEMISTRY OF PLUMBING. 201

alysis has been repeatedly produced by the lead contained in snuff. "In some cases," he says, "death has ensued, and in others serious illness has resulted from the preparations of lead, particularly in the chromate and carbonate of lead used in sugar confectionery, Bath buns, egg and custard powders. The same result has followed the use of wine to which acetate of lead has been added for the purpose of clarifying and sweetening it. Entire districts have been poisoned by lead in cider. Again, at one time—and it is probably still done in some cases —lead was commonly added to the rum in the West Indies." These are cases, and others might be added, in which the true cause of the disease has been traced, and it is no doubt true that causes such as these may be at work in some of the cases supposed to be due to the use of lead pipe. Tanquerel, a celebrated French authority, says that generally the persons who suffer from this disease are those who have to handle the metal or some of its compounds in their business. Of 1213 persons afflicted with lead colic observed personally by this writer, 1050 at least were engaged in operations involving the use of lead or its compounds. This leaves only a small remainder outside of those trades affected with the disease at all. Of this remainder a large number are engaged in occupations which sometimes require the use of a compound of lead. The potters, for instance, use oxide of lead in the glaze which they put on their ware. Now, when it is considered how few are the cases of lead poisoning among those who never use lead in their daily occupations, and also how various are the means by which lead may be introduced into the system, there is left but a small number of cases in which the disease is produced by water drawn from lead pipes.

Pastry and confectionery.
Wine.
Cider.
Jamaica rum.
Tanquerel's observations.

Tanquerel gives the following table of the occupations of men afflicted with lead poisoning who had come under his notice: *Occupation of victims of lead poisoning.*

White and red lead and orange mineral manufacturers................................ 481
Painters.. 390

THE CHEMISTRY OF PLUMBING.

Color grinders	68
Plumbers	14
Platers (in tin and lead)	8
Manufacturers of tin putty	4
Type founders	52
Printers	12
Shot manufacturers	11
Manufacturers of acetate, nitrate and carbonate of lead	10
	1,050
Others	163
Total	1,213

Lead an accumulative poison. Lead is an "accumulative" poison. When taken into the system in small quantities it does not exhibit its effect at once, but as more is taken from time to time, the poison accumulates in the system until enough has been taken to render it operative, and then the person affected suddenly manifests dangerous symptoms. In following out this theory of the accumulative character of lead, it is evident that, no matter how small the dose of lead taken, the occurrence of evil results should be only a matter of time, that is, when the small doses of poison have accumulated to a sufficient amount. *Differences of opinion among physicians.* On this point a difference of opinion has arisen among physicians, some maintaining that the accumulative principle does not hold here, and that, when the lead is conveyed into the system in small doses, it is conveyed out again as fast as it comes in. Others hold that while these small doses are not in themselves able to exert an active poisonous influence, they are the cause of many diseases not usually ascribed to this source. Dr. Muir, previously quoted, states the question thus: "Does this amount of lead thus deposited in the system in any way influence the general health of our bodies? May the healthiness in some places be influenced materially by the amount of lead dissolved by the water in constant use?" *Dr. Dana's opinion.* Dr. Dana thinks "there is reason to believe that a vast number of cases of rheumatic and spasmodic and nervous diseases, a general breaking up, as it were, of the foundations of the great deep of life, have occurred, which can be attributed only to the small daily doses of lead." This is hypothetical. The question is a very important one, but as yet our information is not sufficiently comprehensive or accurate to enable us to answer it.

But while there is not a perfect agreement among physicians as to the amount of lead necessary to produce poisonous effects, it is generally admitted that but little of the metal is required to develop, in time, very serious results. Dr. Parkes, an English authority, Professor Graham and others, think that water which contains one-twentieth of a grain per gallon must be regarded as unsafe. When it is considered that a gallon contains 70,000 grains, it is seen that such a dose amounts to only one part of lead in nearly a million and a half (1,400,000) parts of water. The poisonous character of so small a dose is due to the accumulative nature of the poison. Small as this quantity is, there are those who would fix the limit of safety at even a lower standard. Dr. Adams, of Waltham, Mass., reports a case of poisoning in which only one-hundredth of a grain per gallon was found in the water, and states that such cases are not rare. This would amount to only one part of lead for every seven million parts of water. This statement was made by Dr. Adams in the report of the committee appointed by the American Medical Association to investigate the action of water on lead pipes and the diseases proceeding therefrom. It is safe to infer from these opinions that whenever hydro-sulphuric acid detects lead in water, the use of such water is likely to prove disastrous to some constitutions. *Amount of lead required to exert a poisonous influence.*

It should always be considered in any discussion as to the poisonous dose of lead, that there is a great variation in the susceptibility of various persons to the effects of lead. A case is recorded in which two members of a family were made seriously ill from the use of water containing only, at times, a mere trace of lead—"a quantity," says our authority, "so infini- *Differences in degree of susceptibility to lead poisoning.*

tesimally small as not to have the least effect on the health of the others." It is probable, also, that when once the disease has been contracted by a person he is more susceptible to it than before.

Symptoms and characteristics of lead poisoning. The indications of lead disease are usually pain, constipation, a yellowish complexion, not affecting the eyes or coloring the urine as in ordinary jaundice, and the blue or slatish colored line on the gums. This line is usually located on that portion of the gum that overlaps a tooth. It sometimes happens that this blue line is seen only on a portion of the gum.

The curative treatment of lead diseases has been a subject of much study among medical men. It is too complex a topic to admit of any consideration here. For information on this point the curious reader is referred to the various standard medical works in which the subject is treated.

We now come to the consideration of the manner in which water acts upon the other metals employed more or less extensively in plumbing work. I feel that no apology is needed for having devoted so much space to the subject of lead, as it is the metal most used in pipes for the conveyance of water. Iron, zinc, tin and copper will be considered more briefly.

Iron as a material for water pipes. Were it not for the inconvenience sometimes attending the discoloration of water, iron would possess many advantages over any other metal as a material for water pipes. The purification of water by contact with iron, is a fact well known in chemistry. Prof. Medlock proved by analyses, several years ago, that iron by its action on nitrogenous organic matter produces nitrous acid, which Muspratt has called "nature's scavenger." *Muspratt's and Medlock's experiments* The last-mentioned chemist found, as a general result, that by allowing water to remain in contact with a large surface of iron for about 48 hours, every trace of organic matter was either destroyed or rendered insoluble, in which state it could be removed effectually by filtration. Medlock found, on examining the water at Amsterdam, which smelled and tasted badly, that the sediment charred on ignition and was almost consumed, showing that it consisted of organic matter. He also found that instead of taking iron from the service pipes, the water, before entering those and an iron reservoir, contains nearly half a grain of iron to the gallon; while in the water issuing from the pipes there was only an unweighable trace. Before entering the reservoir, the water holding iron in solution formed no deposits; while the water coming from the pipes, and freed from iron, gave organic sediment above mentioned. He then made analyses of water brought in contact with iron, and water not in contact, with the result that the water which had not touched iron contained 2·10 grains of organic matter, and 0·96 grain iron; the other gave only a slight trace of both, showing plainly that the organic matter in the water was either decomposed or thrown down in contact with iron; and this water when filtered was found to be clear, of good taste, with no smell and free from organic matter. It is not stated in what shape the iron was held in solution, but it was probably in that of carbonate, the usual iron salt of springs, since carbonic acid is so common in water in general. These facts may be made useful under certain circumstances in effecting the purification of water rendered offensive or unwholesome by the presence of organic matter.

Pure water has no action on iron whatever, provided it is free from air or other dissolved gases, especially carbonic acid gas. On the other hand, dry oxygen and dry carbonic acid are unable to attack iron, but in solution they produce the well-known form of oxidation called rusting. That it is the oxygen of the air, and not that of the water, which combines with the iron, can be easily proved by a simple experiment. Take a piece of clear ice, melt it and heat to boiling; after boiling a short time, pour it into a small vial containing some pieces of bright iron wire. The vial must be quite full and tightly corked. Place a similar piece of wire in an open vessel and partially cover it with water. Set both vessels aside for a few days, when it will be found that the wire in the former is still bright, while that in the latter is rusted.

Chemical action of water on iron.

Iron dissolved by carbonic acid. Carbonic acid gas in solution not only attacks metallic iron, but also dissolves it, forming a protocarbonate of iron. If a small quantity of finely divided iron be introduced into a syphon used for transporting mineral water, and the apparatus filled with carbonic acid water under pressure, the iron will soon disappear, being entirely dissolved. This method has been proposed for administering iron medicinally.

Carbonic acid in water. As most kinds of potable water contain either air or carbonic acid in solution, it is evident why iron pipes are attacked by running water. In limestone districts the water seldom contains free carbonic acid, but in every case, unless very impure or just taken from a lively spring, air is present. It has, in fact, been laid down as a rule that no water is fit to drink unless a fish will live in it; and fish cannot live in water that does not contain dissolved oxygen. There are, however, springs in which fish cannot live, but still the water is not unfit to drink, its only fault being a lack of oxygen, which it soon acquires on standing.

Influence of salts. The presence of salts in solution are not without influence when air has access to it; of these, common salt, or chloride of sodium, hastens the rusting, and carbonated alkalies retard it.

Protective oxidation of iron. When a crust of the hydrated oxide of iron has formed on the surface of the iron, it seems to protect the iron by preventing the oxygen from obtaining access to it. This explains the fact that water from new iron pipes contains more iron at first than it does after being in use awhile. Some persons take the trouble to pour thin milk of lime through the pipes and then expose them to the air until it is converted into a dry crust of carbonate of lime, which is a very good protection from rust. Unfortunately, in this as in many other methods of protecting pipes, the sudden jars occasioned by quickly shutting off the water with a full head on, break off this crust.

Protection of a film of oxide. The protection afforded by a film of oxide is well shown by an experiment described in the *Berg- und Huettenmaennische Zeitung* (1873, p. 19). Several pieces of bright wire, some of

THE CHEMISTRY OF PLUMBING. 207

which were protected by a bit of zinc fused on, the others unprotected, were placed in a jar of moist carbonic acid and air; beside them was placed a third lot of pieces of wire which had been heated throughout to a blue shade. The unprotected bright wires rusted in less than 24 hours; those with zinc attached remained free from rust from 3 to 5 days; the blued wires were unattacked for 3 weeks, showing that a blue film of oxide is more effective than contact with an electro-positive metal.

The usual tests for iron is ferrocyanide of potassium, known as yellow prussiate of potash, which produces a deep blue color in dilute solution. Cornelly has even proposed to determine the quantity of iron present by the comparison of the blue colors produced by adding to a solution of ferrocyanide of potassium, in one case a solution of iron of known strength, and in the other the water in which the iron is to be determined. A more delicate test is the sulphocyanide of potassium, which is said to produce a red color when one part oxide of iron in 64,000 parts of water is present. Dollfus states that salicylic acid produces a violet color with one part of sesquioxide of iron in 572,000 parts of water. *Tests for iron in water.*

Zinc is a metal which should never be allowed to come in contact with water which is to be used for drinking or cooking. I make this statement with a knowledge of the wide diversity of opinion which exists among chemists and physicians on this point. Its physical properties as a metal are, I think, very accurately described by Prof. H. von Fehling in *Handwerkerbuch der Reinen und Angewandten Chemie* (IX, p. 899), as follows: "In using zinc for technical purposes it must be remembered that it expands and contracts greatly by change of temperature, and that in cold weather it is especially brittle; and, further, that in contact with other metals, as iron, copper, &c., it readily oxidizes; that it also oxidizes easily in contact with water alone, with brandy, wine, milk and the like, and that the salts are poisonous. It remains unchanged only when in contact with pure olive oil. Sheet zinc must, therefore, have much play so *Zinc not suitable metal for service pipes or tanks. Physical properties of zinc.*

that it may expand or contract. It must be fastened only with zinc nails, or with iron nails thickly covered with zinc. On moist wood it oxidizes very easily. The metal must never be employed for vessels where it can come into contact with food, drinking water and the like."

Chemical action of water on zinc. I take exception to Prof. von Fehling's statement only so far as to claim that zinc is not acted upon by chemically pure water free from air; but the exception is of no practical importance, for the reason that plumbers never have to deal with *pure* water, still less with water free from air or other dissolved gases. It is well known to every one that when a bright surface of zinc is exposed to the air it soon loses its luster from oxidation, the thin film of oxide then formed protecting it from further corrosion. This film adheres so firmly that it can scarcely be removed, and Pettenkofer found the film of oxide on a zinc roof that had been exposed to the weather for 27 years to be only 0·04 inch thick; on a square foot of surface only 142 grains of zinc had oxidized; half of the oxide had been carried off and the other half remained.

Influence of carbonic acid and chlorides on zinc. When zinc is placed in water containing air and carbonic acid, the zinc soon becomes covered with a white coat of basic carbonate of zinc. If the water contains soluble chlorides, such as common salt, it attacks the zinc more violently. Zinrak analyzed a water containing a relatively small amount of chlorides, and found that after standing some time in a zinc reservoir it contained 58·9 grains of zinc in a gallon.

Water from galvanized iron tanks. A French chemist named Roux examined the water kept in galvanized tanks on shipboard and found it turbid; it contained oxide of zinc and suspended particles of carbonate of zinc. These, he remarks, are dissolved by the acids in the stomach and are exceedingly dangerous. The result of Roux's experiments was that the use of galvanized iron tanks in the French navy was forbidden by the war minister.

Prof. Cassell's experiment with galvanized iron. Prof. J. L. Cassels, of the Cleveland (Ohio) Medical College, reported the following interesting experiment made in 1870: " A piece of new galvanized iron chain weighing 1211·95 grains

THE CHEMISTRY OF PLUMBING.

was placed in a glass beaker containing one pint of water taken from the hydrant near the college and loosely covered to exclude dust. In 24 hours the water was of a bluish-white color and tasted distinctly of the salts of zinc. In three days a whitish sediment was observed collecting on the zinc, which was easily detached by agitation. After remaining a week in the water a large deposit of carbonate of zinc was formed, and the water was strongly impregnated with chloride of zinc. Traces of lead were also detected in the water, derived, probably, from the lead impurities in the zinc. The links of chain had decreased 1·04 grain in weight and were heavily coated with the carbonates of zinc and iron.

A similar experiment was recently made with commercial sheet zinc at Columbia College, New York. A strip of zinc weighing 2·22 grams was placed in a gill of Croton water. In a short time it became covered with a white film, a greater part of which fell away on the slightest agitation. The loss of weight in a week was 0·006 gram, or nearly 0·3 per cent. In distilled water it was still greater, or about 0·5 per cent. *Experiment at Columbia College.*

From the foregoing experiments it is evident that zinc, even when alone, is corroded and dissolved by spring, well and river waters without exception. The experiments are of such a nature that any person can repeat them and remove all doubts that may remain in his own mind. The galvanic action which takes place when zinc is in contact with iron or other metals hastens the solution of the zinc, rendering galvanized iron pipes more objectionable than those of zinc alone. Another danger attendant on the use of zinc is the fact that it often contains other and still more objectionable metals, especially arsenic and lead. The difficulty of obtaining zinc free from arsenic is shown in the fact that such zinc sells for 60 cents per pound, whereas at the time of this writing ordinary spelter is quoted at 7½ cents. *All waters corrode zinc. Galvanic action between zinc and iron. Impurities in zinc.*

Zinrak recommends that zinc tanks, when used for water, should be painted on the inside with ocher or asphalt varnish.

210   THE CHEMISTRY OF PLUMBING.

Amount of zinc required to effect a poisonous influence. It seems to be at present a disputed question how much zinc is required to produce serious consequences. According to the United States Dispensatory, "the compounds of zinc are poisonous, but not to the same extent as those of lead. The oxide of zinc used in painting is said to be capable of producing a colic resembling that caused by lead, and called zinc colic." The sulphate known as white vitriol is used externally as a caustic; internally it is tonic, astringent and, in large doses, a prompt emetic. The dose, as a tonic, is 1 to 2 grains; as an emetic, 10 to 30 grains. In an overdose it acts as an irritant poison. Chloride of zinc also acts as a caustic. Internally it is given in doses of a half to one grain; in overdoses it is also a corrosive poison. The oxide of zinc is sometimes administered as a tonic in doses of 2 to 8 grains or more, repeated several times a day.

Zinc compounds.

Prof. Nichols' experiment. Prof. J. R. Nichols states that he examined a whitish powder alleged to have been taken from the joints in the galvanized pipes and found it to consist of carbonate of zinc mixed with a little sesquioxide of iron. In one instance nearly half an ounce of this salt was scraped from the interior of a galvanized pipe 60 feet in extent. The courageous doctor took half a grain of this salt an hour before retiring and passed a very uncomfortable night.

During the past few years a great deal has been communicated to the chemical and medical journals on the subject of zinc poisoning, but the space at command is not sufficient for a review of the testimony.

Tests for zinc in water. Zinc is the most difficult of all the heavy metals to detect, since iron, which is likely to be present in water, helps to conceal zinc. It it safe, however, to predicate in advance that zinc is present if the water has been in contact with that metal. Sulphuretted hydrogen does not precipitate zinc from acid solutions unless acetic acid alone is present. Zinc salts give a white precipitate with sulphide of ammonium and ferrocyanide of potassium. To detect zinc in the presence of iron, add enough ammonia to precipitate the iron and to redissolve the zinc which

THE CHEMISTRY OF PLUMBING. 211

was at first precipitated. Filter and test for zinc in the filtrate by means of sulphide of ammonium or sulphuric acid.

Pure tin is less acted upon, either by water or saline solutions, than any other of the common metals. When exposed to the combined action of water and air, it does, indeed, oxidize slightly, but the oxide being insoluble remains attached to the tin unless mechanically removed. The ordinary constituents of potable water have but little effect upon tin, even when in concentrated solutions. Dilute acids destroy it, even the vegetable acids, as do the caustic alkalies. At the writer's request Mr. E. J. Hallock, of Columbia College, New York, made some interesting experiments to determine the action of saline solutions upon tin, the results of which may be briefly stated as follows: When a piece of block tin, free from lead, is exposed for four weeks to the action of a strong solution of common table salt (chloride of sodium), the solution becomes slightly milky and gives a reaction for tin, although very faint and slowly produced. On filtering, the liquid failed to give any reaction, indicating that the oxide of tin was suspended and not dissolved in the liquid. Strangely enough, the amount of tin in solution at the end of ten months was little, if any, greater than at the end of one month. Several other salts were tried with similar results. Nitrate of ammonia, chloride of magnesium and chloride of calcium acted upon tin sufficiently to give a tin reaction within a few days. Croton water which had been concentrated until it contained 22 grains of salt in a gallon, in contact with tin was soon found to contain a trace of tin. Solutions of chloride of ammonium and of bicarbonate of lime required at least six weeks to acquire a perceptible trace of tin. Sulphate of lime forms a protecting incrustation upon tin.

The nitrates and nitrites have a perceptible action on tin when concentrated. It was not found practicable to determine the loss of weight in the tin, owing to the difficulty of removing the incrustation of oxide formed upon it. Two points were, however, clearly demonstrated: First, that tin is acted upon by most saline solutions, although very slightly, even when exposed for a long time; secondly, the oxide and whatever other compounds—probably oxychloride—were formed remain *suspended* in the liquid and can readily be removed by filtration. Tin salts are not injurious when taken internally; hence, from a sanitary point of view, it is immaterial whether potable water takes up tin from the pipes or not.

Action of saline solutions on alloys of tin and lead. It is a curious fact that saline solutions dissolve out the lead from tin-lead alloys, even if the amount of lead be very small. Weber analyzed a slimy deposit found in a salt-water bath in Reischaur's laboratory and found it to consist of 68 per cent. oxide of tin and 21 of oxide of lead, although the alloy of which the bath was constructed contained but $15\frac{1}{2}$ per cent. of lead to 81 of tin.

Corrosion of tin in well water. It is reported that certain well waters corrode tin pipes rapidly, but I am not able to say to which constituent they owe this corrosive action. It is not impossible that the metal alloyed with the tin in the manufacture of such pipe constitutes a very important factor in the reaction. A well water which is competent to destroy pure tin should, we think, be subjected to chemical analysis before venturing to use it for drinking, the probability being that it contains some unwholesome constituent to the presence of which it owes its corrosive action.

The salts of tin not poisonous. As the salts of tin are not poisonous, their detection is of interest only for the purpose of ascertaining whether a given water is attacking the tin pipes through which it passes, for water containing chlorides and nitrites will generally do so.

Tests for tin. *Chloride of gold*, which can be obtained of any photographer, will produce a purple in very dilute solutions of tin salts. A little nitric acid should be added to the gold solution, and if no purple color appears on mixing it with the water to be tested, it should be allowed to stand a few days, when the purple precipitate will have settled at the lowest point of the test tube, where it is readily seen on placing the tube on a sheet of white paper.

THE CHEMISTRY OF PLUMBING.

Sulphuretted hydrogen yields a precipitate with tin salts, which may be brown or yellow, according to which oxide is present. This precipitate is soluble in alkaline sulphides, and, as above stated, is not formed by sulphide of ammonium.

Copper has far less affinity for oxygen than iron, and will not decompose water except at a bright red heat, if at all. Whether it will under any circumstances is, I believe, a matter of dispute among authorities. Even water which contains acids does not attack copper unless air is also present. On the other hand, in dry air copper is not affected, but air and moisture combined attack it rapidly, especially if any acid, however weak, such as carbonic or acetic acid, be present. Inasmuch as moist air always contains, practically, some carbonic acid, bright copper exposed to its action becomes covered with a film of basic carbonate of copper, very generally but improperly called verdigris. *Action of water on copper.*

Copper not only dissolves readily in the weakest acids, but also in alkaline and saline solutions if exposed to the air. Kersting states that while all potable water dissolves more or less copper from copper pipe or vessels used to hold or conduct it, it attacks the copper with especial violence if nitrate of ammonia is present—a fact which also holds good with regard to tin and lead. If water which holds copper salts in solution is passed through lead pipes, the copper, being more strongly electro-negative than lead, is precipitated by it, and a corresponding quantity of lead is probably dissolved. The same is true to a still greater degree with regard to zinc, so that when water containing copper salts is passed through galvanized iron pipes, the latter are especially attacked by them. This weak galvanic current is well known, and the principle is often employed to protect copper by means of zinc, as in the case of the coppered bottoms of ships and other vessels navigating salt water. The galvanic action of lead and copper is weaker and was long overlooked by practical men, and even yet is not so generally understood as it deserves to be. It is possible that in *Action of acids on copper. Nitrate of ammonia. Galvanic action between copper and other metals.*

cases where pipes have been corroded and no cause could be detected, this weak galvanic action has been at work between the positive pipe and some more negative constituent of the water.

Salts of copper. The carbonate of copper is one of its most insoluble salts, and hence could easily be removed by filtration, because it is merely suspended, not dissolved, in the water. It is said to be decomposed by boiling, forming other insoluble compounds, chiefly the black oxide. The nitrate, sulphate and chloride of copper, which are liable to be produced by the action of waters containing nitrates, sulphates and chlorides, are mostly soluble and possess very dangerous qualities. The true verdigris, or acetate of copper, when brought into contact with water breaks up into two other acetates, one of which contains less copper and is very soluble; the other contains more and is entirely insoluble. It is necessary to remark that this compound is only produced when vinegar or acetic acid comes in contact with copper. It is this which imparts the beautiful green color to cucumber pickles when prepared in copper vessels, as is easily proved by inserting a bright steel knife blade in a green pickle. In a few hours the blade is more or less perfectly copper plated. The oxychloride of copper, which is usually produced when copper is left in contact with salt water or other solution of chlorides, is not soluble in water. It was manufactured and used as a pigment under the name of Brunswick green, but has now given place to the more dangerous Paris green.

Danger of copper utensils in culinary operations. Copper utensils are much employed in culinary operations, and with less danger than would seem at first thought. Boiling water contains no air, and hence, if it contains salts or acids, is not able to attack the copper so long as it continues to boil. Food cooked in copper vessels should not, however, be left therein until cool. The black oxide of copper is soluble in oils and fats, so that greasy matters boiled in copper utensils which are not kept bright are liable to become impregnated with the metal. Considering the risk, their use should be entirely abandoned. Copper salts are highly poisonous, causing vomiting, violent pains in the stomach and bowels, fainting, violent headache, cramps, convulsions and death. *Poisonous character of copper salts.*

In the foregoing pages I have attempted to show, with as much particularity as seemed to be necessary under the circumstances, what results we may expect will follow the exposure of lead, iron, zinc, tin and copper surfaces to the action of water. If the facts are as stated, it is obvious that the only metals which can be counted safe under all circumstances are iron and tin, but these cannot always be used with advantage for economic reasons. Under some—and perhaps many—conditions, lead can be used with safety; but it is well to be sure of our conditions before we trust lead. It is unnecessary, however, to add any general remarks to the very full discussion which has occupied so many pages. *Conclusions.*

CHAPTER IX.

ELEMENTARY HYDRAULICS APPLICABLE TO PLUMBING WORK.

Relation of hydraulics to plumbing.
Plumbers in cities are rarely called upon to face difficulties of a nature requiring a more extensive knowledge of the principles of hydraulics than they may be supposed to have gained in the practice of their trades. All their calculations are based upon definite data. They know the head of water with which they have to deal, and the size and weight of pipe required. They have their constant supply in the street main, and to tap this, bring the water into the house and distribute it, calls for very little of the knowledge which country plumbers must In cities. have to compass equally satisfactory results. All of the science of hydraulic engineering which the city plumber needs to know might be given in a few simple rules and tables; the country plumber, who must often seek his water supply where he can find it, and sometimes bring it long distances through In country. small pipes, must be something of an engineer as well. He certainly meets with difficulties which would puzzle hydraulic engineers accustomed only to large undertakings, such as the construction of water works and the supplying of towns. For the benefit of this large and important class of artisans, as well as of those who employ them, I will briefly consider what seems to me the most important of the elementary facts pertaining to Elementary character of chapter. the science of hydraulics. Were this chapter intended for the perusal of engineers, or those presumably well acquainted with the principles of hydraulic engineering, I should omit many things to which I have given place, and put in many which are here omitted; as it is, my aim is simply to give practical plumbers and others who may be interested in the subject the items of information which my experience has led me to believe they will find most useful. Under the circumstances,

HYDRAULICS OF PLUMBING.

therefore, no apology is needed for the elementary character of this chapter.

Water is a practically incompressible liquid, weighing, at the average temperature of 60° Fahr., about 62·3 lbs. to the cubic foot, and 8·3 lbs. to the gallon. These figures are subject to slight variations incident to changes in temperature. *Water.*

A column of water 12 inches high exerts a downward pressure of about ·43 lb. to the square inch. A column 2 feet high exerts a pressure of about ·86 lb., or just twice that exerted by a column one foot high. This pressure per square inch, due to head, is irrespective of volume or anything else, except vertical hight of column. With these figures in mind, the calculation of the pressure per square inch due to any head is a simple matter. The following rules will be found valuable for reference: *Pressure due to head.*

To find pressure in lbs. per square inch exerted by a column of water.—Multiply the hight of the column in feet by ·43. *To calculate pressure and head.*

To find the head.—Multiply the pressure in lbs. per square inch by 2·31.

Pressure of water.—The weight of water or of other liquids is as the quantity, but the pressure exerted is as the vertical hight. *Weight vs. pressure.*

Fluids press equally in all directions; hence, any vessel or conduit containing a fluid, sustains a pressure on the bottom equal to as many times the weight of the column of greatest hight of that fluid as the area of the vessel is to the sectional area of the column. *Pressure of fluids exerted equally in all directions.*

Lateral pressure.—The lateral pressure of a fluid on the sides of the vessel or conduit in which it is contained is equal to the product of the length multiplied by half the square of the depth, and by the weight of the fluid in cubic unit of dimensions. The following formula is simple and satisfactory: Multiply the submerged area in inches by the pressure due to one-half the depth. By submerged area is meant the surface upon which the water presses. For example, to find the lateral *Lateral pressure.*

218 HYDRAULICS OF PLUMBING.

pressure upon the sides of a tank 12 ft. long by 12 ft. deep; 144 × 144 = 20,736 inches of side. The pressure at the bottom will be 12 × ·43 = 5·16 lbs., while the pressure at the top is 0, giving us, say, 2·6 lbs. as the average. Therefore, 20,736 × 2·6 = 53,914 lbs.

Discharge of water. *Discharge of water.*—The quantity of water discharged during a given time from a given orifice, under different heads, is nearly as the square roots of the corresponding hights of the water in the reservoir or containing vessel above the surface of the orifice.

Relation of discharge to size of orifice. Small orifices, on account of friction, discharge proportionately less than those which are larger and of the same shape under the same pressure.

Circular apertures. Circular apertures are the most efficacious, having less surface in proportion to area than any other form.

Discharge from pipes. If a cylindrical horizontal tube through which water is discharged be of greater length than its diameter, the discharge is much increased. It can be lengthened with advantage to four times the diameter of the orifice.

Contents of pipes. *To find the number of U. S. gallons contained in a foot of pipe of any diameter.*—Square the diameter of the pipe in inches, and multiply the square by ·0408.

Velocity of flow of water. *Velocity of flow of water.*—Water which has a chance to flow downward does so with a velocity in exact proportion to its head. The following table gives the velocity of flow of water due to heads of from 1 to 40 feet:

HYDRAULICS OF PLUMBING. 219

Velocity in Feet per Second due to Heads of from 1 to 40 Feet.*

Head.	Velocity.	Head.	Velocity.	Head.	Velocity.	Head.	Velocity
0.5	5.67	10.5	25.98	20.5	36.31	30.5	44.29
1.	8.02	11.	26.60	21.	36.75	31.	44.65
1.5	9.82	11.5	27.19	21.5	37.18	31.5	45.01
2.	11.34	12.	27.78	22.	37.61	32.	45.37
2.5	12.68	12.5	28.35	22.5	38.04	32.5	45.72
3.	13.89	13.	28.91	23.	38.46	33.	46.07
3.5	15.	13.5	29 46	23.5	38.88	33.5	46.42
4.	16.04	14.	30.00	24.	39.29	34.	46.76
4.5	17.01	14.5	30.54	24.5	39.69	34.5	47.10
5.	17.93	15.	31.06	25.	40.10	35.	47.44
5.5	18.81	15.5	31.57	25.5	40.50	35.5	47.78
6.	19.64	16.	32.08	26.	40.89	36.	48.12
6.5	20.44	16.5	32.58	26.5	41.28	36.5	48.45
7.	21.22	17.	33.06	27.	41.67	37.	48.78
7.5	21.96	17.5	33.55	27.5	42.05	37.5	49.11
8.	22.68	18.	34.02	28.	42.44	38.	49.44
8.5	23.38	18.5	34.49	28.5	42.81	38.5	49.76
9.	24.06	19.	34.96	29.	43.19	39.	50.08
9.5	24.72	19.5	35.41	29.5	43.56	39.5	50.40
10.	25.36	20.	35.86	30.	43.92	40.	50.72

In plumbing work we cannot secure this velocity in the flow of water through pipes because of the friction which constantly tends to diminish it. The longer the pipe the greater the friction and consequent retardation of the flow. In the following table we have the head of water consumed by friction in pipes one yard long and from 1 to 4 inches in diameter. This table shows the head of water required to produce a given flow per minute. By means of the rules given on page 221 it is made applicable to any length of pipe, and a variety of problems relating to lengths and diameters of pipe, discharge in gallons and head in feet are solved by it: *Friction of pipes.*

* Box's Hydraulics.

Head of Water Consumed by Friction in Pipes one Yard Long.*

Loss of head by friction.

Gallons per Minute.	Diameter of the Pipe in Inches.						
	1	1½	2	2½	3	3½	4
	Head of Water in Feet.						
1	.0041	.00054	.00012	.000042	.000016	.0000078	.000004
2	.0164	.00216	.00051	.000168	.000067	.0000313	.0000016
3	.0370	.00487	.00115	.000379	.000152	.0000705	.000036
4	.0658	.00867	.00205	.000674	.000271	.000125	.0000064
5	.1028	.01354	.00321	.001053	.000423	.000195	.000100
6	.1481	.01950	.00463	.001517	.000609	.000282	.000144
7	.2016	.02655	.00630	.002064	.000830	.000383	.000196
8	.2633	.03468	.00823	.002696	.001084	.000501	.000257
9	.3333	.04389	.01041	.003413	.001372	.000634	.000325
10	.411	.0541	.01286	.00421	.00169	.000783	.000401
20	1.64	.2167	.0514	.01685	.00677	.00313	.00160
30	3.70	.4877	.115	.03792	.0152	.00707	.00361
40	6.58	.8670	.205	.06742	.0271	.01253	.00643
50	10.28	1.35	.321	.1053	.0423	.01958	.01004
60	14.81	1.95	.463	.1517	.0609	.02820	.01446
70	20.16	2.65	.630	.2064	.0830	.03839	.01969
80	26.33	3.46	.823	.2696	.1084	.05014	.02572
90	33.33	4.38	1.041	.3413	.1372	.06346	.03255
100	41.1	5.4	1.28	.421	.169	.078	.0401
110	49.7	6.5	1.55	.509	.205	.094	.0486
120	59.2	7.8	1.85	.606	.243	.112	.0578
130	69.5	9.1	2.17	.712	.286	.132	.0679
140	80.6	10.6	2.52	.825	.332	.153	.0788
150	92.5	12.1	2.89	.948	.381	.176	.0904
160	105.3	13.8	3.29	1.078	.433	.200	.1028
170	118.9	15.6	3.71	1.217	.485	.226	.1161
180	133.3	17.5	4.16	1.365	.549	.253	.1312
190	148.5	19.5	4.64	1.521	.611	.282	.1450
200	164.6	21.6	5.14	1.685	.677	.313	.1607
210	181.4	23.8	5.67	1.858	.747	.345	.1772
220	199.1	26.2	6.22	2.039	.819	.379	.1945
230	217.6	28.6	6.80	2.229	.896	.414	.2126
240	237.0	31.2	7.40	2.427	.975	.451	.2314
250	257.1	33.8	8.03	1.633	1.058	.489	.2511
260	278.1	36.6	8.69	2.848	1.145	.529	.2716
270	299.9	39.5	9.37	3.071	1.234	.571	.2929
280	322.6	42.4	10.08	3.303	1.328	.614	.3150
290	346.0	45.5	10.81	3.544	1.424	.658	.3379
300	370.3	48.7	11.58	3.792	1.524	.705	.3617
310	395.4	52.0	12.35	4.049	1.627	.752	.3162
320	421.3	55.5	13.16	4.315	1.734	.802	.4115
330	448.1	59.0	14.00	4.589	1.844	.853	.4376
340	475.6	62.6	14.87	4.871	1.958	.905	.4645
350	504.0	66.3	15.75	5.162	2.075	.959	.4923
360	533.3	70.2	16.66	5.461	2.196	1.015	.5248
370	563.3	74.1	17.60	5.769	2.336	1.072	.5502
380	594.2	78.2	18.57	6.085	2.446	1.131	.5803
390	625.8	82.4	19.56	6.408	2.576	1.191	.6112
400	658.4	86.7	20.57	6.742	2.710	1.253	.6430
410	691.7	91.0	21.61	7.083	2.847	1.317	.6755
420	725.8	95.5	22.68	7.433	2.988	1.382	.7089
430	760.8	100.1	23.8	7.79	3.13	1.448	.743
440	796.6	104.9	24.8	8.15	3.27	1.516	.778
450	833.2	109.7	26.0	8.53	3.43	1.586	.813
460	870.7	114.6	27.2	8.91	3.58	1.657	.850
470	909.0	119.7	28.4	9.30	3.74	1.730	.887
480	948.0	124.8	29.6	9.70	3.90	1.805	.925
490	988.0	130.1	30.8	10.11	4.06	1.881	.964
500	1028.7	135.4	32.1	10.53	4.23	1.958	1.004

*Box's Hydraulics.

HYDRAULICS OF PLUMBING.

The practical application of this table will be found in the following rules:

To find the head of water when diameter and length of pipe and number of gallons discharged per minute are known. — In the above table the head due to a length of one yard is found opposite the number of gallons. Multiply that number by the given length in yards and we have the required head in feet. Thus, to find the head necessary to deliver 130 gallons per minute by a pipe 4 inches in diameter, 500 yards long: Opposite 130 gallons in the table and under 4 inches in diameter is ·679, which, multiplied by 500, gives 339·5 feet, the head sought. *[To find head.]*

To find the diameter of the pipe when head, length of pipe and the number of gallons discharged per minute are known. — Divide the head of water in feet by the length of the pipe in yards, and the number nearest to this in the table opposite the number of gallons will be found under the required diameter. *[To find diameter of pipe.]*

To find the number of gallons discharged when the head, length of pipe and its diameter are known. — Divide the head of water in feet by the given length in yards, and the nearest number thereto in the table under the diameter will be found opposite the required number of gallons. *[To find discharge of pipe.]*

To find the length when the head, number of gallons per minute and diameter of pipe are known. — Divide the given head by the head for one yard found in the table under the given diameter and opposite the given number of gallons, and the result is the required length. *[To find length of pipe.]*

The actual discharge of pipes is easily calculated with approximate accuracy by Prony's formula. In using this formula, find the discharge in gallons per minute by multiplying the head in inches by the diameter of the pipe in inches, and divide the product by the length of the pipe in inches $\left[\dfrac{H \times d}{L}\right]$. In the following table find the number nearest *[Actual discharge of pipe]*

222   HYDRAULICS OF PLUMBING.

to the quotient thus obtained in the first column, and the discharge in gallons per minute will be found opposite it, under the diameter of the pipe used:

Discharge of Pipes by Prony's Formula.

Prony's formula. $\frac{H \times d}{L}$	Velocity in Feet per Second.	Diameter of the Pipe in Inches.							
		1	1½	2	2½	3	3½	4	5	6
		Gallons Discharged per Minute.								
.00002402	.025	.0511	.1150	.2045	.3196	.4602	.626	.818	1.278	1.841
.00005437	.05	.1022	.2301	.4091	.6392	.9204	1.252	1.636	2.556	3.682
.00009108	.075	.1534	.3450	.6136	.9588	1.381	1.878	2.454	3.834	5.523
.0001341	.100	.2045	.4602	.8182	1.278	1.841	2.504	3.273	5.113	7.363
.0001836	.125	.2556	.5750	1.023	1.598	2.301	3.130	4.090	6.390	9.205
.0002394	.15	.3067	.6900	1.227	1.917	2.761	3.756	4.908	7.668	11.05
.0003016	.175	.3578	.8053	1.432	2.237	3.221	4.382	5.728	8.947	12.83
.0003702	.2	.4090	.9204	1.636	2.557	3.682	5.008	6.546	10.23	14.73
.0004452	.225	.4601	1.035	1.841	2.876	4.142	5.634	7.363	11.50	16.57
.0005266	.25	.5112	1.150	2.045	3.196	4.602	6.260	8.160	12.78	18.41
.0006140	.275	.5624	1.265	2.250	3.515	5.062	6.886	9.000	14.06	20.25
.0007080	.3	.6135	1.381	2.454	3.835	5.522	7.512	9.819	15.34	22.09
.0008087	.325	.6646	1.496	2.659	4.154	5.982	8.138	10.64	16.62	23.93
.0009154	.35	.7157	1.611	2.864	4.474	6.443	8.764	11.46	17.89	25.77
.0010286	.375	.7669	1.726	3.068	4.794	6.903	9.390	12.27	19.17	27.61
.0011480	.4	.8180	1.841	3.273	5.113	7.363	10.02	13.09	20.45	29.45
.001274	.425	.8691	1.955	3.477	5.433	7.823	10.64	13.91	21.73	31.29
.001406	.45	.9202	2.071	3.682	5.757	8.284	11.27	14.73	23.01	33.13
.001545	.475	.9713	2.186	3.886	6.077	8.744	11.89	15.55	24.29	34.97
.001690	.5	1.023	2.301	4.091	6.392	9.204	12.52	16.37	25.37	36.82
.002	.55	1.125	2.531	4.500	7.031	10.12	13.77	18.00	28.12	40.50
.00233	.6	1.227	2.761	4.909	7.670	11.04	15.02	19.64	30.68	44.18
.002693	.65	1.329	2.991	5.318	8.309	11.96	16.28	21.27	33.23	47.86
.003079	.7	1.431	3.221	5.727	8.948	12.88	17.53	22.91	35.79	51.54
.003490	.75	1.533	3.450	6.136	9.588	13.81	18.78	24.54	38.34	55.23
.003926	.8	1.636	3.682	6.544	10.23	14.73	20.03	26.18	40.90	58.90
.004388	.85	1.738	3.912	6.954	10.86	15.65	21.29	27.82	43.46	62.59
.004876	.9	1.841	4.142	7.363	11.51	16.57	22.53	29.46	46.02	66.27
.005928	1.0	2.045	4.602	8.182	12.78	18.41	25.04	32.73	51.13	73.63
.00648	1.05	2.147	4.832	8.591	13.42	19.33	26.29	34.37	53.69	77.31
.00708	1.1	2.249	5.062	9.000	14.06	20.25	27.54	36.00	56.24	80.99
.007691	1.15	2.351	5.292	9.409	14.70	21.15	28.80	37.64	58.80	84.67
.008338	1.2	2.454	5.522	9.818	15.34	22.09	30.05	39.28	61.36	88.36
.009	1.25	2.556	5.753	10.23	15.96	23.01	31.30	40.91	63.91	92.04

Discharge of small pipes. The discharge of small pipes may be calculated with sufficient accuracy for practical purposes from the following convenient table, showing the quantity of water that will flow through a pipe 500 feet long in 24 hours, with a pressure due to a head of 10 feet:

⅜-inch bore.... 576 gallons. ¾-inch bore.... 3,200 gallons.
½-inch "   1,150   "   1-inch   "   6,624   "
⅝-inch "   2,040   "   1¼-inch "   .... 10,000 "

Having determined the pressure due to head with which he has to deal, and the size of the pipe needed to discharge a given quantity in a given time, the plumber must calculate the strength which his pipe must possess to resist this pressure under all conditions. This he need not do with absolute accuracy, for the reason that he must use the pipe he finds in the market; but the strength of the sizes in the market is known, and on the basis of this knowledge he can determine the weight of pipe he requires. In all such calculations, however, there should be a liberal margin for safety. The pipe may corrode, external influences may weaken it, and extraordinary pressures may be brought to bear upon it—as by the sudden closing of a cock, which, owing to the incompressible nature of water, causes it to strike a powerful blow, due to the suddenly arrested momentum of the entire column of water in the pipe. This often bursts pipes which are amply strong to resist a great deal more than the normal pressure to which they are subjected. Other causes also operate to increase the pressure and tax the resisting powers of the pipe, and it must be strong enough to bear these without straining. Through the courtesy of Mr. T. O. Leroy, of New York, I am able to present a table of much value, which gives the relation of size and thickness to strength in standard lead pipes. These figures, from which I have omitted the decimals, are compiled from the results of careful tests:

Weight and Strength of Lead Pipes.

Strength of lead pipes.

Caliber	Mark	Weight per foot	Exterior Diam'r	Thickness	Distention on Proof	Absolute Bursting Pressure	Mean Bursting Pressure	Safe Working Pressure	Caliber	Mark	Weight per foot	Exterior Diam'r	Thickness	Distention on Proof	Absolute Bursting Pressure	Mean Bursting Pressure	Safe Working Pressure
⅜	AAA	1 12	.75	.18	.03	1987	1968	492	½	AA	4 8	1.46	.23	.18	950	910	227
⅜	AAA	1 12	.75	.18	.03	1950			½	A	4 0	1.42	.21	.16	810		
⅜	AA	1 5	.68	.15	.07	1610	1627	406	½	A	4 0	1.42	.21	.08	905	857	214
⅜	AA	1 5	.68	.15	.05	1645			½	B	3 4	1.34	.17	.11	790		
⅜	A	1 2	.64	.13	.05	1350	1381	347	½	B	3 4	1.34	.17	.18	700	745	186
⅜	A	1 2	.64	.13	.07	1412			½	C	2 8	1.28	.14	.16	560		
⅜	B	1 0	.625	.125	.03	1330	1342	335	½	C	2 8	1.28	.14	.15	565	562	140
⅜	B	1 0	.625	.125	.03	1355			½	D	2 4	1.25	.125	.14	525		
⅜	C	0 14	.60	.11	.06	1212	1187	296	½	D	2 4	1.25	.125	.18	512	518	129
⅜	C	0 14	.60	.11	.05	1162			½	E	2 0	1.20	.10	.17	475		
⅜	....	0 10	.55	.087	.07	1080	1085	271	½	E	2 0	1.20	.10	.14	475	475	118
⅜	....	0 10	.55	.087	.05	1090			½	....	1 8	1.18	.09	.20	320		
7-16	....	0 9½	.5975	.08	.04	740	775	193	½	....	1 8	1.18	.09	.19	330	325	81
7-16	....	0 9½	.5975	.08	.05	770			½	AAA	6 12	1.80	.275	.20	937		
½	AAA	3 0	1.	.25	.03	1750	1787	446	½	AAA	6 12	1.80	.275	.18	987	962	240
½	AAA	3 0	1.	.25	.08	1825			½	AA	5 12	1.75	.25	.07	885		
½	AA	2 8	.95	.225	.09	1620	1655	413	½	AA	5 12	1.75	.25	.18	762	823	205
½	AA	2 8	.95	.225	.09	1690			½	A	4 11	1.67	.21	.12	690		
½	A	2 0	.86	.18	.07	1425	1393	348	½	A	4 11	1.67	.21	.09	680	685	171
½	A	2 0	.86	.18	.12	1362			½	B	3 11	1.59	.17	.12	505		
½	A	1 10	.82	.16	.06	1230	1285	321	½	B	3 11	1.59	.17	.14	587	546	136
½	A	1 10	.82	.16	.03	1340			½	C	3 0	1.52	.135	.14	415		
½	B	1 3	.75	.125	.06	930	980	245	½	C	3 0	1.52	.135	.15	425	420	105
½	B	1 3	.75	.125	.04	1030			½	D	2 8	1.50	.125	.15	375		
½	C	1 0	.70	.10	.09	790	782	195	½	D	2 8	1.50	.125	.19	325	350	87
½	C	1 0	.70	.10	.07	775			½	....	2 0	1.44	.095		325		
½	D	0 9	.65	.065	.07	462	468	117	½	....	2 0	1.44	.095	.11	320	322	80
½	D	0 9	.65	.065	.06	475			½	AAA	8 0	2.08	.29	.20	730		
½	....	0 10	.65	.07	.09	550	556	139	½	AAA	8 0	2.08	.29	.14	755	742	185
½	....	0 10	.65	.07	.09	562			½	....	7 0	2.	.25	.16	700		
½	....	0 12	.68	.09	.05	637	625	156	½	....	7 0	2.	.25	.16	700	700	175
½	....	0 12	.68	.09	.05	613			½	A	6 4	1.96	.22	.22	595		
½	AAA	3 8	1.10	.23	.14	1510	1548	387	½	A	6 4	1.96	.22	.15	662	628	157
½	AAA	3 8	1.10	.23	.13	1587			½	B	5 0	1.86	.18	.20	500		
½	AA	3 12	1.06	.21	.12	1340	1380	345	½	B	5 0	1.86	.18	.19	512	506	126
½	AA	3 12	1.06	.21	.10	1420			½	C	4 4	1.80	.15	.24	445		
½	A	2 8	1.	.18	.09	1115	1152	288	½	C	4 4	1.80	.15	.20	415	430	107
½	A	2 8	1.	.18	.12	1190			½	D	3 8	1.78	.14	.21	310		
½	B	2 0	.95	.16	.09	1000	987	246	½	D	3 8	1.78	.14	.23	320	315	78
½	B	2 0	.95	.16	.08	975			½	....	3 0	1.74	.12	.34	260		
½	C	1 7	.86	.117	.11	785	795	198	½	....	3 0	1.74	.12	.28	230	245	61
½	C	1 7	.86	.117	.07	805			½	B	5 0						116
½	D	1 4	.84	.10	.09	680	708	177	½	C	4 0	2.04					
½	D	1 4	.84	.10	.08	737			½	D	3 10	2.	.125	.23	325		93
½	AAA	4 14	1.33	.29	.12	1450	1462	365	½	D	3 10	2.	.125	.14	312	318	70
½	AAA	4 14	1.33	.29	.08	1475			½	AAA	10 11	2.60	.30	.15	610		
½	AA	3 8	1.20	.225	.10	1200	1225	306	½	AAA	10 11	2.60	.30	.13	612	611	152
½	AA	3 8	1.20	.225	.07	1250			½	AA	8 14	2.50	.25	.25	512		
½	A	3 0	1.13	.19	.10	1145	1072	268	½	AA	8 14	2.50	.25	.15	510	511	127
½	A	3 0	1.13	.19	.12	1000			½	A	7 0	2.42	.21	.14	405		
½	B	2 3	1.05	.15	.06	890	865	216	½	A	7 0	2.42	.21	.26	405	405	101
½	B	2 3	1.05	.15	.10	840			½	B	6 0	2.36	.19	.27	330		
½	C	1 12	1.	.125	.12	790	782	195	½	B	6 0	2.38	.19	.16	390	360	90
½	C	1 12	1.	.125	.08	775			½	C	5 0	2.32	.16	.13	275		
½	D	1 3	.93	.09	.12	505	505	126	½	C	5 0	2.32	.16	.08	245	260	65
½	D	1 3	.93	.09	.12	505			½	D	4 0	2.18	.09	.22	200		
1	AAA	6 0	1.69	.30	.07	1240	1230	307	½	D	4 0	2.18	.09		200	200	50
1	AA	4 8	1.46	.23	.25	870											

HYDRAULICS OF PLUMBING.

Tin-lined lead pipe is somewhat lighter than lead pipe bearing the same mark, as will be seen from a comparison of the following table with the one last given:

Weights per foot of Tin-lined Lead Pipes.

Caliber.	AAA Weight per ft.		AA Weight per ft.		A Weight per ft.		B Weight per ft.		C Weight per ft.		D Weight per ft.		D Light Weight per ft.		E Weight per ft.		E Light Weight per ft.	
Inches.	lb.	oz.	lb.	oz.	lb.	oz.	lb.	oz.	lb.	oz.	lb.	oz.	lb.	oz.	lb.	oz.	lb.	oz.
⅜	1	8	1	5	1	2	1	0	0	13	0	10			0	8		
½	3	0	2	0	1	12	1	4	1	0	0	13			0	11	0	9
⅝	3	8	2	12	2	8	2	0	1	12	1	8	1	4	1	0	0	12
¾	4	8	3	8	3	0	2	4	2	0	1	12	1	8	1	4	1	0
1	6	0	4	12	4	0	3	4	2	8	2	0			1	8		
1¼	6	12	5	12	4	12	3	12	3	0	2	8			2	0		
1½	9	0	8	0	6	4	5	0	4	4	3	8			3	4		
2	10	12	9	0	7	0	6	0	5	4	4	0						

The strength of tin-lined pipe is about the same as that of lead pipe, the greater strength of the tin being offset by the lighter weight per foot of the pipe thus made. Some experiments made by Mr. A. W. Craven, C. E., chief engineer of the Croton Aqueduct Department of New York, gave the following results:

Size of Pipe.		Lead.	Tin-lined.
A, Breaking strain, per square inch		1500	1600
AA, " " " "		1600	1665
AAA, " " " "		1800	1930

Were the tin-lined pipes made the same weight per foot as lead they would no doubt be considerably stronger. As it is they are probably fully as strong, but I should not advise subjecting them to a greater working pressure than would be considered safe with a lead pipe of the same size and mark. The manufacturers do not claim for it any greater strength than they have allowed for by making the pipe lighter than lead.

Block-tin pipe is stronger for a given weight per foot than lead or tin-lined. As compared with lead its strength is about as 3½ to 1. The following table shows the

Weights per foot of Block-Tin Pipes.

¼ inch,	AA	3½ oz.	⅝ inch,	AAA	11 oz.	1 inch,	A	12 oz.
5-16 "	AAA	6½ "	⅝ "	AA	9 "	1¼ "	AAA	26 "
⅜ "	AAA	7 "	⅝ "	A	6 "	1¼ "	AA	18 "
⅜ "	AA	4 "	¾ "	AAA	13 "	1½ "	AAA	36 "
½ "	AAA	10 "	¾ "	AA	11 "	1½ "	AA	24 "
½ "	....	8 "	¾ "	A	8 "	1½ "	A	19 "
½ "	AA	6½ "	1 "	AAA	17 "	2 "	AAA	40 "
½ "	A	4½ "	1 "	AA	14 "	2 "	AA	26 "

Wrought-iron pipes. Wrought-iron pipes suitable for water service range in diameter from ½ inch to 16 inches. The following table, compiled by Messrs. Tasker & Co., of the Pascal Iron Works, Philadelphia, gives the

Standard Sizes and Weights of Welded Iron Pipes.

Inside Diameter.	Actual Outside Diameter.	Thickness.	Actual Inside Diameter.	Weight per foot of length.
Inches.	Inches.	Inches.	Inches.	Lbs.
½	0.84	0.109	0.623	0.845
¾	1.05	0.113	0.824	1.126
1	1.315	0.134	1.048	1.670
1¼	1.66	0.140	1.380	2.258
1½	1.9	0.145	1.611	2.694
2	2.375	0.154	2.067	3.667
2½	2.875	0.204	2.468	5.773
3	3.5	0.217	3.067	7.547
3½	4.0	0.226	3.548	9.055
4	4.5	0.237	4.026	10.728
4½	5.	0.247	4.508	12.492
5	5.563	0.259	5.045	14.564
6	6.625	0.280	6.065	18.767
7	7.625	0.301	7.023	23.410
8	8.625	0.322	7.982	28.348
9	9.688	0.344	9.001	34.077
10	10.75	0.366	10.019	40.641

HYDRAULICS OF PLUMBING.

These pipes are subjected by the makers to the following tests: *Strength of wrought-iron pipes.*

½ to 1¼ in. butt welded 300 lbs. hydraulic pressure per square inch.
1½ to 10 in. lap   "   500   "   "   "   "   "   "

Practically, they are strong enough to bear any pressure with which the plumber has to deal. The same is true of drawn brass and copper pipes.

The pressures to be dealt with in American plumbing practice vary through a wide range. In cities supplied by what are known as gravity works, *i. e.*, where dependence is placed on natural head at the distributing reservoir, as in New York, the pressure of water is often very light. Where pumping machinery is used and a high head is maintained in tall stand pipes or the pumps deliver directly into the mains, we sometimes get pressures of 100 pounds to the inch and upward. In such cases pipes are subjected to very severe strains, and are often burst by the sudden closing of a cock or valve. The power exerted by a column of water suddenly arrested is almost always great enough to make the pipes quiver and rattle from end to end, and for safety, as well as for convenience, we must make some provision for cushioning the water hammer. An air chamber, always useful, becomes a necessity when heavy pressures are dealt with. These are commonly made by carrying the pipe from 15 to 20 inches above the cock or valve, and in this added length the air is compressed, making an elastic cushion upon which the blow expends itself harmlessly. This, at least, is the theory of the air chamber. Now in practice, where there are heavy water pressures, these air chambers often fail precisely as they sometimes fail in steam fire engines. The air in them is gradually carried out with the water which surges up into them, and when they are full of water, the water hammer, as it is called, is not cushioned, owing to the incompressibility of water. With the hydraulic ram, the fire engine, the steam pump and other hydraulic apparatus, this difficulty is overcome by the use of a small valve, through

Pressures.

Water hammer.

Air chambers

228   HYDRAULICS OF PLUMBING.

which, at certain points in the stroke, air is drawn in in sufficient quantity to supply any loss in the air chamber. In house plumbing we cannot well have recourse to this expedient, and for this reason the common form of air chamber is not usually worth the lead used in making it. An air chamber or its equivalent can, however, be made which will meet all the requirements of the case, and cushion the water hammer what-
Rubber balls. ever the pressure. Take a piece of iron pipe, say 7 or 8 inches long, and large enough to hold a couple of solid rubber balls of $2\frac{1}{2}$ or 3 inches diameter, such as can be procured at any toy store. Cover the top of the tube with a screw cap, and fit it in position by means of a reducing collar and nipple. A chamber of this kind never fails. The blow of the water is expended in compressing the rubber balls. It would be well if such chambers were attached to the service pipe system of every house, since they not only greatly reduce the strains upon the pipes, couplings and closet valves, but in case of freezing permit such expansion within the pipe as will be very apt to avert the nuisance of a burst. I know of cases in which pipes have been protected by this simple device during repeated freezings, and can recommend it with entire confidence as likely to save a great deal of money in cold weather to landlords and tenants.

Location of air chambers. The location of the air chamber depends upon circumstances. A safe rule is to place it as near as possible to the point where the shock is felt most sharply. The philosophy of this is that the column of water, flowing rapidly and forcibly from a cock or fixture, is suddenly arrested by the closing of the cock, and strikes a blow precisely in the same manner as a stick of timber when used as a battering ram, but with this difference, that the blow is distributed at once through the whole mass of the water in the pipes. The concussion should be arrested at the point where it is produced, which is at the cock, basin, or closet, as the case may be. The resulting jar when a cock is closed is heard all over the house, it is true, but the first strain is felt at the point where the water is shut off.

HYDRAULICS OF PLUMBING.

In providing for the storage of water, the plumber usually makes his calculations by "rule of thumb." Unless guided in this by knowledge gained from long experience, he is very likely to make mistakes. In the case of cisterns for rain water, it is important for economic reasons that he should make them large enough to contain all the water which will drain into them; for sanitary reasons it is desirable that they should not be unnecessarily large. The following rules and information will be found to apply especially to cisterns: *[margin: Storage of water. Cisterns.]*

To calculate the amount of water which will drain from a roof.—Multiply the area of the roof in feet by the average rainfall in a month, in inches, and the product by ·623. This gives the number of gallons which will drain from the roof in a month. *[margin: Drainage of roofs.]*

With a regular consumption for domestic purposes, cistern capacity for one-quarter to three-eighths this amount of water will be ample.

When a roof has a steep pitch, its size should be determined by the area of ground it actually covers.

The amount of rainfall in a given locality is determined by averages. The average rainfall in vertical inches is 30 inches in the basin of the great lakes; 32 inches on Lake Erie and Champlain; 36 inches on the Hudson River, at the head waters of the Ohio, through the central portions of Pennsylvania and Virginia, and the Western portion of North Carolina; 40 inches in the extreme Eastern and Western portions of Maine, Northern New Hampshire and Vermont, Southeastern Massachusetts, Central New York, Northeastern Pennsylvania, Southeastern New Jersey, and Delaware; also in a narrow belt running down from the Western portion of Maryland, through Virginia and North Carolina, to the Northwestern portion of South Carolina; thence up through the Western portion of Virginia, Northeast Ohio, Northern Indiana and Illinois to Prairie du Chien; 42 inches on the East coast of Maine, Eastern Massachusetts, Rhode Island and Connecticut, and middle part of *[margin: Rainfall in United States.]*

Maryland; thence on a narrow belt to South Carolina; thence up through Eastern Tennessee, through Central Ohio, Indiana and Illinois to Iowa; thence down through Western Missouri and Texas to the Gulf of Mexico; 45 inches from Concord, N. H., through Worcester, Mass., Western Connecticut and the City of New York to the Susquehanna River, north of Maryland; also at Richmond, Va.; Raleigh, N. C.; Augusta, Ga.; Knoxville, Tenn.; Indianapolis, Ind.; Springfield, Ill.; St. Louis, Mo.; thence through Western Arkansas, across Red River to the Gulf of Mexico. From the belt just described the rainfall increases inland and southward until, at Mobile, Ala., it is 63 inches. The same amount also falls in the extreme southern portion of Florida.

Obstructions in pipes. An important subject intimately related to practical hydraulics, is the obstruction of pipes from causes other than the complete or partial closing of the water-way by the lodgment *Air traps.* of solid substances therein. It is a matter of frequent experience in plumbing work that water cannot be made to flow through pipes under certain conditions frequently met with, until openings are made to allow the air confined in the pipes to escape. A great many cases of this kind have been brought to my notice during the past few years by correspondents in different parts of the country, and examination has usually shown that the trouble resulted in bends of the pipe. It often happens that, through carelessness, or because of difficulties in the way of laying them straight, pipes have high and low *High and low points.* points. The latter give little trouble, unless by collecting and holding sediment; but when we have low points we are very certain to find high points, and when these occur they are likely to give trouble. In Figure 22 is shown an exaggeration of one of these high points, or upward bends, which will serve for purposes of illustration. Water always contains more or

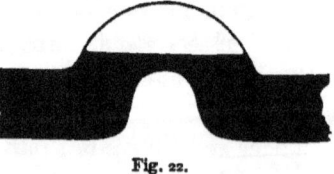

Fig. 22.

HYDRAULICS OF PLUMBING.

less air and undissolved gases, which, being lighter than water, naturally collect at the highest points in a long pipe or conduit, and, as shown in the illustration, partially close the waterway. When several of these obstructions occur in a line of pipe, it is not unusual for the flow of water to be stopped altogether. If the pressure of water was exerted only on one side of such an air cushion, it would be quickly dislodged, but in the case shown in the figure the pressure is the same above and below the bend, consequently the water only presses upward. Obstructions of this kind give a vast amount of trouble to plumbers, and as the principle involved is not generally understood, a few facts on the subject which I take from a paper by Mr. Richard H. Buel, C. E., will be of interest. I have changed the numbers of his drawings to make the figures number in regular order: "The collection of air in high points of a pipe may stop the flow of water altogether, even when a considerable pressure is applied. In Fig. 23 (1) represents a bent tube containing a liquid, which stands at the same hight (on the level of the line $a\ b$) in each leg when no pressure is applied. Suppose

R. H. Buel

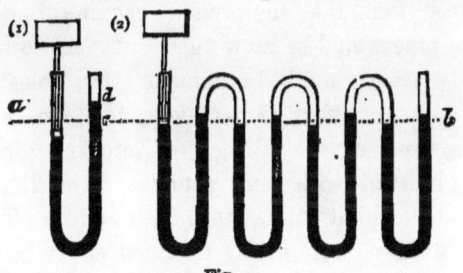

Fig. 23.

Illustration of action of air in water pipes.

that, by means of a weight and piston, a pressure is put upon the liquid in one leg, producing a rise of $c\ d$ in the other. In the same figure (2) represents a set of four bent tubes connected together, each similar to (1), and with the same amount of liquid in each, the upper connections being filled with a lighter liquid. If the same pressure is applied to this set, it will only cause a rise of about one-fourth of the amount that occurred in (1), since there are four equal columns to resist the pressure. This principle is applied to great advantage in the construction of mercury gauges, obviating the necessity of employing a high column to register a considerable pressure of steam. But it will be seen that the application of this principle is anything but desirable in the case of water pipes. Thus suppose that the pressure applied to one leg of (1) was just sufficient to force water out of the other leg; then, if the pipes were arranged as in (2), about four times as much pressure would have to be applied to one open leg to force water out of the other. An arrangement something like this may occur in a long water pipe, being represented on an exaggerated scale in Fig. 24. We

Fig. 24.

may suppose that this pipe was at first full of air, and that water, being admitted, compressed the air into the high points, and collected in the low points, very much as in (2) Fig. 23. It might happen, then, that the pressure required to force water through the pipe would be more than sufficient to burst it; so that the only remedy would be to make a few holes and release the air. Such an effect is not very common, only occurring in the case of very long and crooked pipes. It is frequently observed with long syphons when the accumulation of air at the high points stops their action. The remedy is to have valves at such points that can either be opened by hand from time to time to let out the air, or can be arranged so as to open automatically when the air attains a certain pressure.

Obstructions in city mains. This trouble is sometimes experienced in the mains of city water works, and it is attended with great danger if the water is forced through the mains by an engine, since, if a pipe becomes air bound, the effect may be to increase the water pressure, and thus burst the pipe. In laying a city main, it is often impossible to avoid high points on account of the nature of the ground; but they should be carefully noted by the engineer, and at every such point means should be provided for drawing off the air. If possible, it is a good plan to place a hydrant at each high point, as it will probably be used often enough to allow the air to escape. Wherever the main is tapped for the purpose of furnishing water to a building, it is well to drive the tap into the upper part of the main, so that it will aid in relieving the pipe of air. In pumping water through a main after the engine has been stopped for some time, it is necessary to see that the valves at the principal high points are kept open until the pipe becomes filled with water."

The only correction I consider it necessary to make in Mr. Buel's explanation of the causes of, and remedy for, the interruption of the flow of water in pipes, relates to what seems to be a misuse of the word syphon. A syphon, strictly speaking, is a tube through which water passes, partly by gravity and partly by the atmospheric pressure, or suction, so called. In other words, the water is lifted above the level of the water in the reservoir by means of the column of water in the longer leg, which, by its weight, is constantly tending to form a vacuum in the highest part of the pipe, into which vacuum water from the reservoir is forced by atmospheric pressure. Opening the top of a syphon of this character would merely empty both legs of the pipe. If, however, the top of the bend in the pipe be below the level of the water in the reservoir, the opening of the pipe at the bend allows the air to escape. This form of pipe is often called a syphon, but incorrectly. The pipe on the left of the vessel containing water shown in Fig. 25 is a syphon, while that on the right is not. From a syphon the air can be drawn by means of a pump, but not by simply opening a valve, it being necessary to create a partial vacuum in order to remove the air. It is quite common to speak of any vertical bend in a pipe as forming a syphon, but under ordinary circumstances syphons are rarely formed in pipes, since they can only

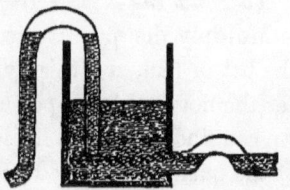

Fig. 25.

Syphons.

Difference between a syphon and a bend.

be made to lift water about 33 feet at the sea level when working to the best advantage. They work under the same conditions in this respect as a suction pump.

Raising water. In practical plumbing it is not always possible to deal with natural heads. Water must often be raised—as from wells and cisterns, and when drawn from sources which give an available head but not strength enough to carry them to the level required—by mechanical means; and the problem is how to raise *Pumps.* most water with the least expenditure of power. When no head is available, recourse must be had to pumps. These may be driven by wind, by water under certain circumstances, by steam or caloric engines, by animal power or by hand. The following tables will be found useful for theoretical calculations:

To Find the Power Required to Raise Water to any Hight.—Multiply the quantity required per minute in cubic feet by the lift in feet, and this by 6·23, dividing the product by 33,000 for the nominal horse-power required. By adding 30 per cent. we can find the actual horse-power required in most instances. If the quantity required per minute is in gallons, the multiplier will be 8·3 instead of 62·3.

Table of the Power Required to Raise Water from Deep Wells.

Diameter of pump barrel.	Description of pump.	Quantity of water raised per hour.	Maximum depth from which this quantity can be raised by each unit of power.			
			One man turning a crank.	One donkey working a gin.	One horse working a gin.	One horse-power steam engine
Inch.		Gallons.	Feet.	Feet.	Feet.	Feet.
2	Double action lift and force pump.	225	80	160	560	880
2½...........		360	50	100	350	550
3		520	35	70	245	385
3½...........		700	25	50	175	275
4		900	20	40	140	220

HYDRAULICS OF PLUMBING.

The problem of raising water was the first one of a mechanical or engineering character which mankind was called upon to solve. The most barbarous races as well as the civilized are alike compelled to draw water. The means used may be simple or complex, but the necessity is equal in both cases and, in not a few instances, the mechanism is identical. The question of how to raise water is not only the most important but the most frequently recurring of all the mechanical problems which the modern engineer has to solve, and, unlike most other engineering problems, this is a question which personally concerns every individual of the community. The earliest device for the purpose was probably an earthen pot or a bag of skin attached to a cord and let down to the spring or into the well.

Primitive methods of raising water.

The shadoof, or common well sweep, seems to have been the next step, and from drawings found in Egypt it is proved that this device is at least 3000 years old, and probably even older. Of simple forms of water-raising contrivances, such as flutter wheels, chain pumps, Persian wheels—having a number of pots upon a rope or chain—and the simple suction pump, it may be safely said that there is little, if anything, new for the last thousand years or more, modern progress consisting chiefly in improvement in workmanship, better materials and a greater attention to the details.

Early forms of pumps, &c

Very early in the history of the world animal power was used to assist in the raising of water, and tread wheels, horizontal winding drums and the direct attachment of animals to the bucket rope, which was led over a pulley, were some of the more common means used. The plumber, in dealing with the question of water raising, has usually to depend upon manual labor or upon some motor, as wind, steam or hot-air engines. Animal power is rarely employed, because a "horse-power" or similar machine for utilizing the force of animals usually costs more than a small steam engine or other prime mover of equal power.

Motive power

Men taken at an average are equal to the production of one-

Hand power.

fifth of a horse-power for ten hours per day. A strong man has, for a few minutes at a time, exerted a force equal to more than half a horse-power, lifting a weight of 18,000 pounds one foot high in a minute, but this could not be kept up. In estimating the quantity of water required to be raised, a man's power can be estimated as equal to the raising of 5000 pounds one foot high per minute. In putting in a pump to be worked by hand, a mistake is often made in choosing one in which the leverage is so large that the hand does not have a perceptible resistance, and is obliged to travel over a very great distance to do the work. The books give a resistance of 30 pounds and a speed of $2\frac{1}{2}$ feet per second as the greatest rate of speed at which work can be kept up. The weight, I should judge, was about right, but it seems to me that the speed is much greater than can be conveniently maintained in pumping. I should think that a double 18-inch stroke would be much nearer a practicable rate. That is 3 feet per second, but only half of the time performing work.

When a man has to work a pump for a short lift, we see no objection to the use of a good-sized barrel so as to obtain a fair amount of resistance. This reduces the time necessary for pumping a given quantity of water, though it makes the work a little harder. Where a pump has to be used by women and children, especially if the whole distance through which the water is carried is considerable, a pump which works easily is absolutely necessary. In such cases a pump with a long leverage and a comparatively small bore must be selected. For a well or cistern from which a great deal of water is to be drawn by different persons—as, for instance, one by which a large school is supplied—it is necessary that the pump should deliver a large quantity of water at each stroke. No one individual pumps more than one or two pailfuls at a time, and it makes little difference whether the whole force is expended in two or three strokes or in seven or eight. It makes a vast saving in time, however, when the pail is filled in two strokes. In

HYDRAULICS OF PLUMBING. 237

setting up a pump that delivers a great quantity of water at each stroke, care should be taken to have a large nozzle and a free water-way, otherwise the stream will be too violent and spatter and splash. This is a very common fault with many pumps when they are worked rapidly.

The distance to which water can be raised by the common lifting pump varies with the hight above the sea level and also with the pressure of the atmosphere. At the sea level the column of water that the atmosphere will support is about 33 feet in hight, and a pump will draw water, as it is called, this distance; but it must be remembered that the force which sends the water into the pump at this hight is so small as to be almost balanced by the weight of the water; hence a lifting pump would deliver water very slowly drawing it this distance. The nearer the pump barrel is to the surface of the water, the more rapidly the pressure of the atmosphere forces the water through the suction pipe. Hence many pump makers, in putting up a pipe, never put it further than 25 feet from the water level. This sends the water to the pump with a force nearly equal to a head of 7 or 8 feet. Where a greater distance is unavoidable—as, for example, where the suction pipe must be very long and the pump has a large bore and is worked rapidly—a vacuum chamber is very valuable in preventing the water from "breaking" in the pipe. With pitcher and other pumps having very large cylinders, the suction pipe can rarely be made large enough to supply the pump, and when working fast there is a loss both of power and capacity. My experience with pumps leads me to think that a vacuum chamber is very desirable at any time. I have seen a pump of, say, 2 or 2½ inch bore supplied through a long half-inch pipe fitted with a vacuum chamber, and found that by the most rapid pumping it was almost impossible to "break" the water in the suction pipe.

High to which water can be raised by atmospheric pressure.

Vacuum chambers.

In putting up pumps plumbers frequently pay too little attention to details. When a pump is ordered care should be taken to obtain one suitable for the work to be done, neither too

Setting up pumps.

large nor too small, and the connections should by all means be properly made. Not long since a boiler pump was returned to a manufacturer because it would not work, and on examination it was found that the suction pipe had been put on to the delivery opening and the delivery pipe on the suction. It was of course little wonder that the engineer could not get his boiler full of water. It often happens that a house pump is put up in such a way that the water cannot be made to run down. This may happen through accident or design. Where, on the approach of cold weather, the plumber carelessly leaves the house pump in such a condition that the water cannot be made to run out of the pipe, he should be held responsible for the damage resulting. Such carelessness should always be discountenanced, as it brings the trade into bad repute.

Wooden pumps.   Until within a few years the form of pump in common use consisted of a single log of wood bored out and provided with a spear, two valves and a spout. The bark was removed, but there was seldom any attempt to shape the log or reduce its size, unless, perhaps, around the top. The objection to this form of pump was found in the fact that the wood decayed and the inside of the pump barrel disintegrated. The surface of the wood also became slimy, and after a few years' use the water would be found charged with particles of wood fiber and fungoid growths. Their durability was surprising, however, and in spite of the objections named, water was delivered by them in very pure condition—at least until the pumps had become old. The selection of the log determined in a great degree the life of the pump. But while in some respects admirably adapted to outdoor wells of moderate depth, they were not efficient in delivering water from wells of 60 feet or more in depth, as the power required to work them was out of all proportion to the amount of water raised. The reason for this was the necessarily large size of the bore and consequent heavy load always on the plunger. While still in limited use, however, wooden pumps of this kind have been to a great

HYDRAULICS OF PLUMBING. 239

extent superseded by lighter and cheaper ones made by machinery.

Following the primitive form of wood pump came the chain pump, which was also adapted to raising water from wells of moderate depth. This, although one of the oldest forms of pumps known, has come into use in this country within twenty years. It is very cheap, simple, durable, and will rarely freeze in the coldest climate. A chain pump will raise water with great rapidity—faster, perhaps, from wells of moderate depth than any other mechanical device in use. In deep wells, however, the labor of raising water by the chain pump is very severe, as there is a long column of water to be supported and the leakage is considerable. The waste of power increases as the tubing wears, giving the chain free play from side to side. The only really objectionable feature of this pump is the zinc coating which it is commonly considered necessary to give the chain. The chances of zinc poisoning from this cause are very small, but I have heard of instances in which zinc poisoning has been traced to it and proved by crucial tests. *Chain pumps.*

Since the day of the chain pump the iron pump has come into more general use than any other device for lifting water. What is commonly known as the cistern pump is made by all pump manufacturers and has become standard. In their general features all pumps of this class are alike, consisting of a cast-iron cylinder with spout, and base for securing it to the platform upon which it stands; a brake and its fulcrum, or stand; a piston, piston rod and valves. These pumps are in use in nearly all parts of the world, and have been for some years an important article of export. In this country they are used by the million, and, all things considered, they are the cheapest, most durable and most efficient hand pumps ever made. In these pumps the diameter of bore ranges from 2 to $3\frac{1}{2}$ inches, increasing by quarters of an inch. The pipes used with them are from $\frac{3}{4}$ inch to $2\frac{1}{4}$ inches, and may be of any kind known to the trade. The following table shows the average efficiency *Iron pumps. Cistern pumps.*

HYDRAULICS OF PLUMBING.

Duty of cistern pumps. of good pumps of this pattern, worked moderately with one hand:

Diameter of bore.	Gallons per minute.
2 inch	6
$2\frac{1}{4}$ "	8
$2\frac{1}{2}$ "	12
$2\frac{3}{4}$ "	15
3 "	22
$3\frac{1}{4}$ "	26
$3\frac{1}{2}$ "	30

Proper sizes of pipes. The size of pipes used with pumps of this class should be determined with reference to the hight to which the water has to be rasied. The following table will be useful to those who put in pumps and make the connections:

Size of bore.			Size of pipe.
2 inch.	For any ordinary hight		$\frac{3}{4}$ inch.
$2\frac{1}{4}$ "	Under 18 feet		$\frac{3}{4}$ "
	Over 18 feet		1 "
$2\frac{1}{2}$ "	Under 18 feet		1 "
	Over 18 feet		$1\frac{1}{4}$ "
$2\frac{3}{4}$ "	Under 18 feet		$1\frac{1}{4}$ "
	Over 18 feet		$1\frac{1}{3}$ "
3 "	Under 18 feet		$1\frac{1}{3}$ "
	Over 18 feet		$1\frac{3}{4}$ "
$3\frac{1}{4}$ "	Under 18 feet		$1\frac{3}{4}$ "
	Over 18 feet		2 "
$3\frac{1}{2}$ "	Under 18 feet		2 "
	Over 18 feet		$2\frac{1}{4}$ "

Pumps of this class weigh from 15 to about 50 pounds each. Leather valves and packing are commonly used, but brass valves can be had from the makers when hot water is to be pumped. **Durability of cistern pumps.** Properly cared for, these pumps will wear for an indefinite period. Various parts may get out of order, and persons inexperienced in such matters are apt to think that a new pump is needed. Commonly this is a mistake. Pumps of this class are made on the system of interchangeability of parts, and any part which wears out or breaks can be replaced at small cost. The most expensive part of a small size of this style of pump—the cylinder—costs less than half the price of a new pump, and $1·50 will replace all the parts likely to wear out in many years' service. A few cents spent on new leathers as often as may be necessary, and an occasional tightening of screws and nuts, will extend the life of such a pump indefinitely. If a pump "runs down" when left standing for a few minutes and water must be poured into the barrel to make the piston suck, it needs attention. The repairs necessary to correct these defects are easily made, but if neglected the pump will rapidly wear out. <small>Repairs.</small>

For outdoor work iron pumps are rapidly superseding other kinds. One of the prime essentials for an outdoor pump is that the brake shall be long enough and the barrel high enough, so that it may be worked by a person of common hight, standing. When the barrel of the pump is above ground, however, there was always danger of freezing in cold weather, and the first great improvement in this class of pumps consisted in sinking the working parts below the surface. Up to that time pumps of this class had been of the ordinary suction-pump pattern, the water flowing immediately from the piston out of the spout. We now have three classes of these pumps—lift, lift and suction, and suction and force. In the lift pump the barrel and lower valve are carried down below the surface of the water, the upward stroke of the piston carrying up the water raised without the aid of atmospheric pressure. This form of pump is much used, especially in driven wells. The working parts are perfectly protected from frost; they are simple and strong, and may be removed without trouble. The lower cylinder is made very compact in form, so as to go into the bore of a driven well, and is commonly provided with a strainer of some sort, which is screwed upon the end. When the cylinder is not long enough to reach the water level, a length of suction pipe is attached, and the pump then sucks as well as lifts. The capacity <small>Iron pumps for outdoor work. Precautions against frost. Lift pumps.</small>

of such a pump is about the same as that of a cistern pump of the same diameter lifting water the same distance, ranging from 8 to 26 gallons per minute. In the more perfect form it is so arranged as to allow the water to run back when desired, to prevent freezing; the brake-stand swivels so as to make it either a right-hand or a left-hand pump, and by adding to the wrought-iron set-length and piston rod, it is adapted for use in wells of almost any depth. When the well is very deep, however, it may be necessary to increase the leverage by lengthening the brake, and to counterbalance the added weight we have the greater weight of the piston rod and column of water. It is frequently an advantage in deep wells to use a strainer provided with an iron rest, which projects far enough to be firmly imbedded in the earth at the bottom of the well. This holds the pipe steady and aids in supporting it. Pipes in deep wells should be well braced, as the jarring and hammering of the brake is usually great enough to rack a long line of pipe, loosen the connections and necessitate frequent repairs. In light, sandy soil, cisterns, dug wells, and in any situation where there is danger of drawing dirt into the pipe, and where there is room enough to use a large strainer, the so-called "Mushroom" strainer presents many advantages. This strainer is of the saucer shape and the water enters it at the top, while that which runs back from the barrel when the valve is tripped flows out of the strainer in an upward direction, thus preventing the roiling of the water by stirring up the mud and sand on the bottom.

The lift and force pump differs from the lift and the suction and lift pumps in an arrangement of parts by which the water is ejected from the cylinder under pressure great enough to carry it beyond the point at which power is applied. As adapted to ordinary work, force pumps are always piston pumps, arranged with an air chamber to equalize the pressure and afford a constant stream instead of an intermittent one, which, by its action, might seriously strain the pipe. The force pump

HYDRAULICS OF PLUMBING. 243

is the one which in cities is most frequently used, since it is not only able to lift water above the point at which power is applied, but to send it in any direction and to almost any distance. It is largely used for raising water to tanks on the upper floors of houses supplied from mains in which the pressure is not great enough to give the required head. These pumps usually require more power for a given lift than any other, owing to the greater friction of parts. There are one or two hand-force pumps, however, in which the internal friction is but little, if any, greater than in the most efficient lift pumps. The efficiency of the best of these pumps, provided with an air chamber and worked with sufficient power, may be averaged as follows: *Efficiency.*

Size of bore.	Size of pipe.	Gallons per minute.
2 inch	$\frac{3}{4}$ inch	6
$2\frac{1}{4}$ "	$\frac{3}{4}$ "	9
$2\frac{1}{2}$ "	1 "	12
$2\frac{3}{4}$ "	$1\frac{1}{4}$ "	15
3 "	$1\frac{1}{2}$ "	22
$3\frac{1}{2}$ "	2 "	30

The efficiency of a pump without a vacuum chamber will be somewhat less than this, as it might be found difficult under certain circumstances to work the pump to its capacity. The power necessary to obtain this efficiency depends, of course, upon the hight to which the water has to be forced, as well as the distance. When one of the larger sizes is employed for raising water to a great hight, one man would probably be unable to work the pump to its capacity. The force pumps of all leading manufacturers are able to do this amount of work. The amount of power required, of course, depends upon circumstances. When the pump is continually supplied with all the water it can take, the amount of power required will be at a minimum, and the pump will be able to work up to its full capacity. A vacuum chamber on a small suction pipe is almost a necessity, because it frequently happens, in a city, that the *Power required.* *Vacuum chambers.*

small head of water in the street mains, and the small pipe used to bring water, prevent a sufficient supply from reaching the pump, and consequently the pump does not do half the work of which it is theoretically capable. The addition of a vacuum chamber below the pump keeps a constant stream flowing to the pump, and at the same time acts as a reservoir from which the pump may draw a supply at each stroke. The ordinary water charger or primer used on common suction pumps answers this purpose, and adds greatly both to the ease and the capacity of a force pump under the circumstances named. Their cost is small, but their utility is very great and will repay the expense and trouble of applying them.

Primers.

Pumps for city houses. In city houses the pump most used for raising water is a side pump mounted on a plank. These pumps are often sold unmounted. They are very convenient to fasten to the side of a building or partition, as they have side ears, while the suction pipe and lower connection can be got at without disturbing the pump. The brake is usually arranged so as to be right or left hand as may be desired. The parts are commonly all brass. Sizes vary from the small $2\frac{1}{2}$-inch bore, with a capacity of 12 gallons per minute, up to $4\frac{1}{2}$-inch bore, capable of delivering 50 gallons per minute. When a steady and constant stream of water is required to be forced up, and a rapid supply needed, a double acting suction and force pump is used. The pumps deliver water at both upward and downward strokes. A pump with $2\frac{1}{4}$-inch bore will deliver about 16 gallons per minute; with a $2\frac{1}{2}$-inch bore, 24 gallons; $3\frac{1}{4}$-inch bore, 52 gallons; $4\frac{1}{4}$-inch bore, 100 gallons per minute. Such pumps, when furnished with an air chamber and hose, are very effective for throwing a stream of water either for fire purposes or for washing windows and carriages and sprinkling walks. The larger sizes are very heavy and require so much power that a power pump would in many cases be preferable. In putting up pumps of this class large pipes are absolutely necessary, since the waste of power in forcing through small pipes the large quantity of water they deliver, is enormous.

Double-acting suction and force pumps.

HYDRAULICS OF PLUMBING.   245

When as large a quantity of water as these pumps will throw is to be raised by hand-power, some form of pump with a double brake is commonly used, so that two men can work at the same time.

In this country hand pumps are made in almost unlimited variety. Our manufacturers have brought the business to a high standard of excellence, and in no country of the world are pumps made which are so cheap and efficient as ours. The illustrated catalogues of our leading pump manufacturers are so full of exact and specific information that no one who has a correct idea of the work to be done need make a mistake in choosing a pump that will do it. *Variety of hand pumps.*

There is scarcely any work which the laboring man is called upon to do which is more irksome than that of pumping when the labor is to be long continued or when the quantity of water is large. On this account it is always desirable to employ power for pumping where that is possible. In country towns horse-power is frequently available for this work, the so-called "horse-powers," either double or single, being readily arranged to drive a pump. These are not, however, sufficiently common or cheap enough to be very generally available. The best power for driving small pumps, in locations where it can be used, seems to be wind. A small windmill, working, as it does, for a good proportion of the time, is a much more reliable power than is generally believed. *The labor of pumping.*

Wind was one of the first sources of power utilized by man. In Holland, windmills have for a very long period furnished power for grinding, pumping and draining, and in that country the windmill of large size has been brought to a degree of perfection of which we have little idea. Mills of very large size and great power are used, and for a long time were able to compete with steam engines as sources of power, even when a considerable amount was needed. Now, however, the steam engine furnishes power, where a great deal is needed, as cheaply as a windmill. The reason for this is that the large mills cost a *Wind power.*

great deal of money; indeed mills costing from $10,000 to $20,000 are not unfrequently met with. These require even more attention than an engine of the same power, and do not work continuously. For small powers, however, they have a great many advantages, especially when they are to be employed at such work as that of water raising.

Windmills. A good windmill will head itself to the wind from any direction without attention. It governs its own speed, not increasing above what is desired, even in a heavy gale. It can work constantly day and night as long as there is wind. To be durable it must be well built and furnished with self-oiling boxes. It is important to have it noiseless in its action, especially if it is to be located near a dwelling. It is not worth while for any one to attempt to build a windmill, as a much better machine can be bought for less money than it would cost to make one. These mills are usually regulated by an adjustment of the sails or vanes. In the large mills in Holland the canvas which covers the arms is taken in when the wind blows hard, more and more being removed as the force of the wind increases. In this country it is found better to turn the slats or vanes so that the wind has less effect upon them. One of the best means of doing this seems to be to turn the slats edgewise toward the wind, the slats being arranged in frames for this purpose.

Utility of windmills. In the Western States, in level countries, on the tops of lofty hills and along the seacoast, windmills do more work than in sheltered places among hills or in a well-wooded country. Thus in Kansas, California or Texas a windmill will do double the work it will in Central New York. The stronger and more continuous the wind, the more power will be obtained, and a wind blowing 50 feet per second will give four times the power of a wind at 25 feet per second.

Location. In locating a windmill care should be taken to set it in as exposed a situation as possible. This is usually done by placing the mill upon the top of a building on a framework erected for the purpose, and generally directly over the spring or well

HYDRAULICS OF PLUMBING.

from which water is to be taken. There are windmill pumps, however, which work at a distance from the mill. In this case both mill and pump can be located in the places best adapted to them.

The following table shows the average power of windmills of different sizes: *Pumping power of windmills.*

    9 feet mill from 1 man to ½ horse-power.
    12   "   " 2 men to 1½   "
    17   "   " 4 men to 3   "
    25   "   " 1 horse to 6   "
    40   "   " 10 horse to 20   "

Best Diameters for Pumps.

Diameter of Windmill.	Elevation in Feet.											
	10	15	20	25	30	40	50	60	80	100	125	150
8 feet........	5	4	3¾	3½	3¼	3	2½	2¼	2	1¾	1½	1¼
9 "	6	4½	4	3¾	3½	3¼	3	2½	2¼	2	1¾	1½
10 "	6½	5	4½	4	3¾	3½	3¼	3	2½	2¼	2	1¾
12 "	8	7	6	5	4½	4	3¾	3½	3¼	3	2½	2
14 "	9	8	7	6	5	4¾	4½	4¼	4	3½	3	2½
17 "	12	10	9	8	7	6	5	4¾	4½	4	3½	3
25 "	15	13	12	10	9	8	7	6	5½	5	4½	4

Diameter of pumps

The stroke of the pump is assumed to be from 4 to 6 inches; but many mills are so arranged as to allow a variation of the length of stroke according to the force of the wind and amount of work done.

The speed at which mills can be driven varies, of course, *Speed.* with the speed of the wind and the load. The following is an approximate statement of the number of revolutions per minute. Above these velocities the regulators begin to act to prevent any increase:

Diameter.   Revolutions
Feet.   per minute.
   8 75
   9 .... 60 to 70

Diameter. Feet.	Revolutions. per minute.
10	60 to 65
12	50 to 60
14	45 to 50
17	35 to 45
25	20 to 30
40	12 to 15

Construction. The smaller sizes of mills are set upon cast-iron columns or timber frames. The larger sizes are usually placed on the top of rectangular towers formed of four strong timbers set inclining toward each other, and strongly braced to make the whole firm. Where large sizes are necessary, the manufacturers furnish drawings and specifications showing how the framing, &c., must be set up. In this case the manufacturer needs to know the depth of the well or spring below the surface of the ground; the least depth of water ever known in it; the hight above the platform of the well to where the water is discharged; the lateral or side distance (if any) from the supply *Adaptation of windmills to conditions.* to the place where the water is to discharge; the amount or quantity of water wanted, or at least the purpose for which it is to be used; also the extent or quantity of water afforded by the supply or source; and the hight at which the mill must be erected to secure a free current of air. In case of a bored or driven well, he should know the diameter.

There are windmills in New York, built a number of years ago, which pump water into tanks on the tops of lofty buildings at a merely nominal cost—in one case the repairs for some five or six years amounting to but few dollars—the cost for pumping being practically only the interest on the first outlay. *Cost* The prices range from $75 to $80, for the smallest sizes, to something like $2000 for the 40-feet mills. In places where fuel is very costly, it may at times be economical to employ still larger mills, especially if it is a place where strong winds prevail.

In Hingham, Mass., a 9-foot windmill was erected several

years ago which lifts water 50 feet above the pump and forces it 450 feet through an inch pipe. A mill 9 or 10 feet in diameter, when well constructed, seems in most locations to be amply able to lift and force water to an elevation of 60 feet, and supply it in sufficient quantity for a large house with bath rooms, water-closets and the like. In some places windmills have been used for the purpose of drainage. The following is the description of one used for draining after the Dutch plan; it serves to show the power of a comparatively small windmill: The tower is 27 feet high, with a building 22x24 and 14 feet high, used as a house for a family. The whole is built on piles driven into the soft ground. The mill, 25 feet in diameter, drives a baling wheel 11 feet in diameter, 17 inches face, placed in a large wooden box or receiver to which the drains lead. The water is lifted 6 feet, and in an ordinary wind, when the baling wheel makes four to five revolutions per minute, it raises and discharges 1920 to 2400 gallons per minute, and in a strong wind, at seven revolutions, 3360 gallons, or 84 barrels per minute. *Examples of efficient windmill pumps.*

In one instance of which I know, a small windmill takes water 550 feet distant from the house and raises it with ease to an elevation of 65 feet. During a stiff breeze it has pumped 600 gallons in an hour. There are very few days in which the mill cannot work at least some part of the day, and by having ample tank room the supply is always sufficient for lavish use.

In closing these remarks upon windmills, I cannot do better than present the following extract from an article upon the subject in the *American Agriculturist:* "A few years ago a windmill was an unusual sight in this country, except in the very oldest portions. We were not a sufficiently settled people, and did not remain long enough in one place to make it profitable to build such substantial mills as have been so long in use in other countries; we needed cheaper and more quickly constructed mills. Those which we could then procure were not satisfactory; they were slightly built, and were not able to take *Mechanical application of wind power.*

care of themselves when the breeze became a gale or a hurricane. Recently our mechanics have turned their attention to wind engines, and great improvements have been made in their construction. We have now a choice of several kinds of them, all of them useful, but differing chiefly in their degree of adaptation to varying circumstances. At the recent Illinois State Fair there were no less than thirteen different wind engines on exhibition, from the small one, 8 feet in diameter, costing but $100, of but half a horse-power, and fitted for pumping stock water or churning, to those of 30 or 40 horse-power, costing $3000, and able to run a grist mill or a woolen factory. Between these extremes there are a number of mills capable of adaptation to almost every purpose for which power is needed on the farm or in the workshop. A mill 22 feet in diameter, costing about $500, has a power of five horses; a two horse-power mill is about 16 feet in diameter, and costs about $325. The cost is less than that of a steam engine, and a wind engine needs neither fuel nor skilled attendance. Neither is there danger of fire or explosion from accident or carelessness. The wind engines are now made self-regulating, and in a sudden storm close themselves. They are also made to change their position as the wind changes, facing the wind at all times. On the Western prairies, and almost everywhere, except in sheltered valleys in the East, we have wind enough and to spare, which offers to us a power that is practically incalculable and illimitable, and the means of utilizing this power is cheaply given to us in the numerous excellent wind engines now manufactured. In fact, so cheaply can these mills be procured, that it will not pay for any person to spend his time in making one although he may be a sufficiently good mechanic to do it."

Pumping by steam. In a great many locations where power has to be employed in raising water, steam is the only power which can be conveniently applied. It is suitable for almost any situation, is easily managed, is generally understood by mechanics, and presents no difficulties not easily overcome. Its universal adaptability

HYDRAULICS OF PLUMBING.

and the immense demand for steam-driven pumps has turned the attention of engineers and capitalists in this direction, and at the present time the manufacture of steam pumps and their accessories is one of the largest industries in the country. It is interesting to note the fact that James Watt, the so-called father of the steam engine, was really a steam pump man, all his engines for a great many years being devoted entirely to the pumping of water out of mines. The application of the steam engine to the furnishing of power for other purposes was done by other persons while Watt was busy with pumps. The manufacturer of to-day has so simplified and cheapened the steam pump that, while its cost is very small, its management is so simple that it may almost be said to be perfectly automatic. The chief item of cost, and the portion of the apparatus requiring the most attention and care, is the boiler. In cases where steam for heating is employed, a steam pump can be used without any additional trouble. Many people fear to use steam boilers on account of the supposed danger attending them and an idea that the insurance will be increased by them. There are a great number of boilers in the market which can be used in insured buildings, the companies considering them no more dangerous than a coal stove.

I have in mind one among the many excellent steam pumps for light duty which may be taken as an illustration of the best machines of its kind in use. The pump is 2 inches in diameter and six inches stroke. The steam cylinder has the same length of stroke, and is $5\frac{1}{2}$ inches in diameter. The pump discharges ·08 of a gallon of water at each stroke, and when running at an ordinary rate of speed makes 100 strokes and delivers 8 gallons of water per minute. It can with ease be run up to a speed of 150 strokes, when it would deliver 12 gallons per minute. The pump will run even faster than this, but it would not be advisable to keep it running steadily at a higher speed, because the wear and tear would become too great. The boiler, which consists of a coil of steam pipe inclosed in a suitable case, is perfectly safe against explosion. The ordinary pressure carried is from 5 to 30 pounds per square inch, while the boiler is tested to 300 pounds per square inch. With 7 pounds pressure in the boiler the pump will force water 50 feet high. In such cases the boiler is fed from the tank, the pressure being sufficiently great to force the water into it against the pressure

Boiler. of the steam. The boiler and pump form an arrangement complete in itself, and may be used for warming as well as pumping, the boilers in such cases being made larger to suit the service required of them. An indicator or steam gauge is attached which shows the pressure, and there is a safety-valve by which the pressure is prevented from rising above the desired point.

Pump. The pump is so arranged that it is always ready to start as soon as there is steam pressure in the boiler, provided, of course, the steam valve is left open. The amount of coal required to run

Coal consumption. one of these pumps is very small. It is stated on good authority that 30 pounds of coal will run one 8 hours, discharging 13 gallons of water per minute 95 feet high, or a little more than 6200 gallons of water raised 65 feet high at a cost of, say, 10 cents.

Economy. The cost for pumping the same amount by hand would be at least $1·50, and perhaps more. One of these pumps and boilers is calculated to furnish all the water required by 12 families, yet they are capable of doing much more. In a French apartment house in New York one of these pumps and boilers is supplying 24 families with water. It is in this case, however, somewhat overtasked, and the supply at times is a little scant.

Duty. With a pressure of 12 pounds per square inch, pumping 70 feet high, one of these pumps has run continuously at 150 strokes per minute, delivering upward of 700 gallons per hour.

Details. When less water is needed the pump can be run slower and the consumption of coal will be proportionately less. The principal parts of the pump are brass, for the purpose of preventing corrosion. The steam pipe is half inch in diameter and the exhaust three-quarters. The discharge pipe is 1 inch; suction, $1\frac{1}{2}$ inch. The boiler is but 3 feet high and takes up a space

2 feet in diameter. In case of the grossest neglect possible, the only damage which could be done to the boiler by cutting off the supply would be to rupture one of the pipes of which it is made, and so allow steam and water to escape and put out the fire.

Another pump for a similar purpose, but constructed on an entirely different plan, has recently been attracting a good deal of attention. Gas or kerosene is the fuel used. It is not of the direct-acting kind, like that just described, but has an oscillating cylinder which drives after the ordinary manner a shaft to which the pump is attached. This machine is made perfectly automatic in all respects save, perhaps, that of oiling all its bearings. It keeps the steam pressure constant by turning on or off the gas or kerosene as the pressure tends to vary. The water supply is also self-regulating, the feed pump sending water into the tank when it is not needed in the boiler. When gas is used, five or six minutes are sufficient to get up steam. One of these engines will pump 10 barrels of water per hour at a cost of about 6 cents. It only occupies about as much space as a flour barrel, and weighs 250 pounds. This engine possesses another point, sometimes of great value, and that is it can furnish power for light work, like running a turning lathe, sewing machine and the like. The principal objection to these pumps is that they are not very strong and are likely to wear out somewhat sooner than is convenient. *An automatic steam pump.* *Cost of running.*

One of the most common methods of raising water by power is by using the so-called hydraulic ram. The simplicity of operation of the hydraulic ram, its effectiveness and economy, together with the fact that it is applicable in thousands of situations where it is now unknown, render a better knowledge of its operations desirable. The hydraulic ram is decidedly the most important and valuable apparatus yet developed in hydraulics for forcing a portion of a running stream of water to any elevation proportionate to the fall obtained. It is perfectly applicable where not more than 16 inches fall can be had; yet the *Hydraulic rams.* *Where used.*

greater the available head the more powerful the operation of the machine and the higher the water may be conveyed. I know of a ram working near Philadelphia which, with a head of 16 inches, raised 40 feet all the water needed to supply a large farm. It has been in use 25 years. The relative proportions between the water raised and wasted are dependent entirely upon the relative hight of the spring or source of supply above the ram and the elevation to which it is required to be raised—the quantity raised varying in proportion to the hight to which it is conveyed with a given fall. The distance which the water has to be conveyed and consequent length of pipe has also some bearing on the quantity of water raised and discharged by the ram, as the longer the pipe through which the water has to be forced by the machine the greater the friction to be overcome and the more power consumed in the operation; yet it is common to apply the ram for conveying the water distances of 100 and 200 rods, and up elevations of 100 and 200 feet. Ten feet fall from the spring or brook to the ram is abundant for forcing up the water to any elevation under, say, 150 feet in hight above the level of the point where the ram is located; and the same 10-foot fall will raise the water to a much higher point than that last named, although in a diminished quantity in proportion as the hight is increased. When a sufficient volume of water is raised with a given fall it is not advisable to increase the fall, as in so doing the force with which the ram works is increased, the amount of labor which it has to perform greatly augmented, the wear and tear of the machine proportionately increased and its durability lessened; so that economy in the expense of keeping the ram in repair would dictate that no greater fall should be applied for propelling the ram than is sufficient to raise a requisite supply of water to the place of use.

<small>Conditions of economy and efficiency</small>

<small>Calculating the fall required for a given duty.</small> To enable any person to make the calculation as to what fall would be sufficient to apply to the ram to raise a sufficient supply of water to his premises, I would say that in conveying it an ordinary distance of, say, 50 or 60 rods, it may be safely cal-

HYDRAULICS OF PLUMBING.

culated that about one-seventh part of the water can be raised and discharged at an elevation above the ram five times as high as the fall which is conveyed to the ram, or one-fourteenth part can be raised and discharged, say, ten times as high as the fall applied; and so on in proportion as the fall or rise is varied. Thus, if the ram be placed under a head or fall of 5 feet, of every 7 gallons drawn from the spring one gallon may be raised 25 feet or half a gallon 50 feet. Or with 10 feet fall applied to the machine, of every 14 gallons drawn from the spring one gallon may be raised to the hight of 100 feet above the machine.

The following is an example of what a ram will do when properly set up and with supply and other things proportioned to each other. The fall from the surface of the water in the spring is 4 feet. The quantity of water delivered every 10 minutes at the house is 3¼ gallons, and that discharged at the ram 25 gallons. Thus nearly one-seventh of the water is saved. The perpendicular hight of the place of delivery above the ram is 19 feet, say 15 feet above the surface of the spring. The length of the pipe leading from the ram to the house is 190 feet. This pipe has three right angles, rounded by curves. The length of the drive or supply pipe is 60 feet; its inner diameter 1 inch. The depth of water in the spring over the drive pipe is 6 inches. The inner diameter of the pipe conducting the water from the ram to the house is three-eighths of an inch. *Example showing the efficiency of a ram.*

It is essential that the drive or supply pipe should be on the curve of quickest descent to get the full value of the head. This approximates a catenary. If on a regular grade, the bottom water runs away from the top water so to speak. *Supply pipes for rams.*

Care should be taken to set the ram in a pit deep enough to protect it from frost, or else the frost should be kept out by boxing and packing.

The following table gives the capacity of rams of different sizes, together with the weights and diameters of pipes to be used in connection with them: *Capacity of rams.*

HYDRAULICS OF PLUMBING.

Size of Ram.	Quantity of water furnished per minute by the spring or brook to which the ram is adapted.	Length of Pipe.		Caliber of Pipes.	
		Drive.	Discharge.	Drive.	Discharge.
No. 2.	3 quarts to 2 gallons.	25 to 50 feet.	Where desired.	¾ in.	⅜ in.
No. 3.	1½ " " 4 "	" "	" "	1 "	½ "
No. 4.	3 " " 7 "	" "	" "	1¼ "	½ "
No. 5.	7 " " 14 "	" "	" "	2 "	¾ "
No. 6.	12 " " 25 "	" "	" "	2½ "	1 "
No. 7.	20 " " 40 "	" "	" "	2½ 2¾ "	1¼ "
No. 10.	25 " " 75 "	" "	" "	4 "	2 "

Weight of Pipe if of Lead.

Size of Ram.	Drive Pipe for any head or fall not exceeding 10 feet.	Discharge Pipe for not over 50 feet rise.	Discharge Pipe for over 50 and not exceeding 100 feet in hight.
No. 2.	6 pounds per yard.	8 pounds per rod.	14 pounds per rod.
No. 3.	8 " "	11 " "	16 " "
No. 4.	10 " "	11 " "	16 " "
No. 5.	23 " "	20 " "	28 " "
No. 6.	40 " "	6 " per yard.	8 " per yard.
No. 7.	40 to 45 "	9 " "	11 " "
No. 10.	48 lbs. per yard c. iron.	20 " "	23 " "

Supply and discharge pipes. If the ram is to be placed under a greater head or fall than named in the above table, it will of course be necessary to increase the weight and strength of the drive, or supply, pipe; also, if the water is to be forced to any greater hight than above mentioned, the discharge pipe should be proportionately increased in weight and strength. Where the water is to be forced to any great distance (say more than 1200 feet) it is preferable to use a discharge pipe of larger caliber than named in the above table.

Size of ram. With a given supply of water under a great fall, the ram is not required to be of a larger size than for the same quantity of water under a less fall. That is, a No. 4 ram would be of sufficient capacity for taking the water from a spring or brook furnishing 7 gallons per minute where the fall is 8 or 10 feet; if there is not over 3 or 4 feet fall to the same spring or brook, then a No. 5 ram would be better adapted to the place.

HYDRAULICS OF PLUMBING.

If the stream is a large one and a greater supply of water be required than one of the large-sized machines will supply, it is better to increase the number of machines than to increase the size of the one in use. Several rams may be set so as to play into one discharge pipe, each having a separate drive pipe. *Working rams in batteries.*

The durability of rams under constant service is quite wonderful. I know of one put up in Durham, Conn., in 1847, which had been in constant use up to the time when I last heard of it, in 1873. It had not cost $5 for repairs and seemed good for many years more. The drive pipe was 1¼ inch bore, 40 feet long. The discharge pipe was half inch in diameter and 825 feet long. The water was discharged 85 feet above the ram in a perfectly steady, continuous stream. *Durability of rams.*

There are many subjects omitted from this chapter which might properly be considered under the head of elementary hydraulics; but as most of those which seem to me of especial interest in connection with plumbing work are considered more or less fully in other chapters, their omission here is due rather to design than to oversight.

17

CHAPTER X.

SANITARY CONSTRUCTION AND DRAINAGE OF COUNTRY HOUSES.

Health dependent upon good drainage. Health and comfort in country houses depend upon the selection of a well-drained site. If the natural drainage is not good, it must be artificially drained by one of the several approved methods, which need not be described here in detail. A location which cannot be drained should never be chosen, and, as the rule, those which are not naturally well drained are not desirable. *Filled lands.* This is a point which should be very carefully looked after, especially in the suburbs of large towns, where marsh and low lands have been filled in to raise them to the desired grade. In such cases the level of the subsoil water is likely to be dangerously near the surface. Filling in a basin, or low swamp-hole, does not change the level of standing water, and land made over such original depressions, unless exceptionally well underdrained, is almost certain to be an unhealthy site to build upon. *Underdrainage.* The importance of underdraining filled land was very strikingly illustrated during the epidemic of cerebro-spinal *Cerebro spinal meningitis in N. Y.* meningitis in New York during 1872. In the early months of the epidemic, and before the disease spread throughout the more densely populated districts of the city, it was found that in a majority of instances the spread of the infection was along the lines of the old water-courses, long ago filled in and forgotten, clearly showing that the filling up of natural springs and water-courses without providing for the thorough drainage of the soil, is dangerous to public health. Our civil engineers are beginning to understand this better than they did a few years ago, and we are likely to have fewer mistakes of this kind in the future than in the past.

Stagnant water. Surroundings should also be looked after. Stagnant water should not be allowed to remain anywhere in the neighborhood.

Running water rarely remains impure for any length of time, as its organic impurities are gradually oxidized and enter into combinations which render them harmless; but when water stands, as in ponds without outlets, in undrained swamps, &c., it is a fruitful source of malaria. The early morning is the best time in which to choose a site for a country house—supposing, of course, that the person proposing to build is in a position to select an eligible location. If one place is covered with a fog, while other places are free from it, the choice should lie in favor of the latter. The presence of such a fog, or even a thin, opalescent mist, indicates wet ground; and although there may be no appearance of standing water on the surface, the source of the excessive moisture in the air will be found under the surface, if sought. *Morning mists.*

The subject of land drainage has a literature of its own which is so complete that I need not extend the scope of this volume to include it. Those interested in the subject can find several cheap and excellent manuals on land drainage on the shelves of any general bookseller, and the most that I can attempt in this place is to urge the importance of the subject upon all into whose hands this work may pass. The almost universal prevalence of fever and ague attests the need of more thorough drainage of districts in which the value of land is great enough to justify the expenditures needed. There is scarcely a place within forty miles of New York that is free from intermittent and worse fevers, and not one that I have seen which could not be made healthful if the proper means were taken to drain the soil. To secure good results the drainage of a populous district must be undertaken as a public work; but so general is the indifference still manifested to sanitary reform, that it is always difficult to secure the popular consent to the levying of a tax for any such purpose. We shall be wiser in these matters a generation hence. *Land drainage.* *Fever and ague.*

The plan of a house and the direction in which it fronts are not always matters to be determined by the preferences of the *Plan and position of a country house.*

owner. When practicable, however, as is generally the case in isolated country houses, it is desirable to give as many of the living and sleeping rooms as possible the benefit of abundant sunlight. This is usually best secured by giving them a southern exposure. Broad piazzas, heavy vines trained upon trellises, and overhanging shade trees are very attractive and beautiful, and often comfortable during the warm days of summer; but in so far as they exclude the sunlight and render a place "damp," they are bad. We cannot afford to make too many sacrifices to secure picturesque effects, and the differences which the observant traveler notices between our country houses and those of Europe are largely due to differences of climate and other circumstances. Experience has shown that health and comfort are promoted by giving the sunlight a fair chance to penetrate to every nook and corner to which it can make its way. It will do more than tons of disinfectants to purify and sweeten the environments of our dwellings. Human beings are as dependent upon the vitalizing and energizing power of sunlight as are the plants in our conservatories or the vegetables in our kitchen gardens. A house hidden in the deep shadows of great trees and surrounded by broad, curved piazzas, always seems to me like a gloomy man with overhanging brows sitting in the Valley of the Shadow of Death; and I never find myself in such a mansion, even in the hottest of summer weather, without involuntarily recalling the lines:

> "Blest power of sunshine, genial day,
> What balm, what bliss are in thy ray!
> To feel thee is such perfect bliss
> That had the world no joy but this—
> To sit in sunshine, calm and sweet—
> It were a world too exquisite
> For man to leave it for the gloom,
> The dim, cold shadow of the tomb."

Sunshine is rarely appreciated, though it comes to us with blessings woven into every ray; and the sanitarian who should devote a lifetime to proclaiming its benefits would do more to

DRAINAGE OF COUNTRY HOUSES. 261

promote public health than any who have yet entered this wide field of philanthropic labor.

It does not follow, however, as the logical sequence of what has already been said, that the occupants of country houses must altogether dispense with vines and shade trees. These are eminently desirable in their proper places, only we must not let them come between us and the sunshine. The greatest favor that Alexander could do the philosophic Diogenes was to step aside and permit the sunshine to fall into the tub which gave the old cynic shelter. Let us, who boast a larger knowledge and a broader and more comprehensive philosophy, be not less wise than the ancients in matters which concern us so deeply as this. Science has taught us that the sun is the source of all life. All terrestrial phenomena are dependent upon light, heat and actinic force, and when these are excluded life and vigor yield to death and decay. We know how dependent plants and all living organisms are upon the sun, but we are apt to forget that we need the sunshine as much as plants and flowers—vastly more, indeed. *Vines and shade trees.* *Sunshine.*

When health is a consideration—and I do not need to say that health is not always considered—the occupant of a country house should see that his cellar is clean, dry and well ventilated. If possible it should be light, for we are not likely to have any one of the three essential conditions above mentioned in any place where daylight never comes. In a great many instances cellars are allowed to become so foul as to be a perpetual menace to the health of those living over them. When sickness comes how seldom do we look for the cause of it in the right place, if at all. As the rule, country cellars are damp, mouldy vaults, chiefly useful as places for the storage of the winter supplies of vegetables. To suggest putting provisions anywhere else would shock a farmer's sense of propriety; but in all the buildings on his farm he could not find a worse place for the storage of vegetables than the cellar under his house. Many of my readers well know what cleaning out the cellar in the spring *Clean, dry and well ventilated cellars* *Causes of sickness often found in cellars.*

Decomposition. means, and how much decayed and mouldy vegetable matter in advanced stages of decomposition is usually gathered up from the floor. A farmer would be shocked and disgusted if it was suggested that a sheep's carcase be allowed to rot all winter in the cellar; but it is a well-known fact that the danger to health from decaying animal matter is small compared with that resulting from the decay of vegetable substances. A little care expended in keeping the cellar clean would be amply repaid; but unless the broom and shovel are supplemented by abundant fresh air and wholesome sunlight, the labor of purification will never be fully accomplished.

Wet cellars. When from any cause a cellar is liable to be wet, either from the inflow of water under or through the foundations or by soakage through the soil, it should be drained. I have seen cellars which were always dry, and I have known of one in which cider has been kept for 20 years without turning to vinegar, and a buck-saw might lie on the floor for an indefinite period without showing a spot of rust; but such cellars are not common, and an arrangement for drainage should be provided in all but exceptional cases. In his excellent work on "Farm Drainage," published some 20 years ago and still standard, **A New England cellar in spring time.** Judge Henry F. French draws the following vivid picture of a New England cellar in spring time, which is so appropriate to the subject we are considering that I cannot resist the temptation to quote it:

"No child whoever saw a cellar afloat during one of these inundations will ever outgrow the impression. You stand on the cellar stairs, and below is a dark waste of waters of illimitable extent. By the dim glimmer of the dip candle a scene is presented which furnishes a tolerable picture of chaos and old night, but defies all description. Empty dry casks, with cider barrels, wash tubs and boxes, ride triumphantly on the surface, while half-filled vinegar and molasses kegs, like water-logged ships, roll heavily below. Broken boards and planks, old hoops and staves, and barrel heads innumerable, are buoyant with this change of the elements, while floating turnips and apples, with here and there a brilliant cabbage head, gleam in the subterranean firmament like twinkling stars, dimmed by the effulgence of the moon at her full. Magnificent among the lesser vessels of the fleet, like some tall admiral, rides the enormous mashtub, while the astonished rats and mice are splashing about at its base in the dark waters like sailors just washed at midnight from the deck by a heavy sea.

"The lookers-on are filled with various emotions. The farmer sees his thousand bushels of potatoes submerged and devoted to speedy decay; the good wife mourns for her diluted pickles and apple sauce and her drowned firkins of butter, while the boys are anxious to embark, on a raft or in the tubs, on an excursion of pleasure and discovery."

This picture, though drawn with the free hand of caricature, is not greatly exaggerated. I have many times witnessed such a scene, and not a few of my readers will recognize it as something which has come within their own experience. Cellars liable even to excessive dampness, and especially those subject to inundation, are unsafe. The drainage of a cellar can usually be accomplished without difficulty by means of earthen tiles. The methods will be found fully described in any good work on land drainage. <small>Wet cellars unsafe.</small>

A barn and its surroundings may be a perpetual nuisance or not, according to circumstances. Ordinarily it is clean enough inside, but the cattle yard is generally so foul that, except in unusually dry weather, one who ventures to cross it must tread ankle deep in filth of the nastiest description. A neglected pigstye is another horror—disgusting to look at and giving off a pestilent effluvium day and night, to be wafted, with the mingled musk and ammonia odors of the barn-yard, into open windows and doors. Such a disregard of sanitary laws, to say nothing of the violation of decency involved, is without excuse, and its only explanation is found in the charitable supposition of ignorance on the part of those responsible for it. I have <small>Barns and barn-yards. Pig-styes.</small>

seen barns that were as clean in themselves and all their surroundings as the houses of the people owning them. This can
never be when manure is spread out over the barn-yard to rot
in the open air. Everything in the way of manure, including
weeds, fallen leaves, refuse vegetable matter, carcases of dead
animals, kitchen garbage, animal excreta—in fact everything
capable of fermentation and decay—should be composted and
utilized. Not being a farmer, either scientific or practical, I
will not venture specific recommendations as to the best and
most economical methods of composting manure on a large
scale for profit, but a few suggestions on this point may be of
interest to those who, for sanitary reasons, are willing to take
the trouble of making muck heaps for the safe and convenient
disposition of whatever might give rise to nuisance if left to
ferment and decay in its own way. Others are referred to the
several able and exhaustive works on the subject, written by
eminent scientific agriculturists, which may be had of any bookseller.

The theory of composting waste organic matter is to provide for the decay and transformation into useful, or at least
harmless, compounds. The means by which this can be accomplished are numerous and exceedingly simple, entailing no
expense which is not more than offset by the value of the
manure made, and no trouble that is not vastly more than com-
pensated by the sanitary benefits attained. All that is necessary is to thoroughly intermix and cover the matter to be
treated with any light, dry, absorbent substance, and keep it on
a dry bottom under cover. The substances suitable for
covering are dry mould, peat, spent vegetable ashes, marl,
sawdust, crushed straw and many other substances equally
cheap and available. Sand and clay are not suitable. A
superior material for composting may be made by mixing
peat, wood ashes and dry mould. When composting is to be
done on a small scale, the first treatment of the matter to be
composted can be carried on conveniently and safely in a

DRAINAGE OF COUNTRY HOUSES.   265

large box or tank. This may be made the receptacle for everything suitable for transformation into manure, and when full the contents may be removed and piled under a shed until needed for use. If the person who takes the trouble to make a compost for sanitary purposes has no use for the manure, he can usually sell it to those who are intelligent enough to know its value for a good deal more than an equal bulk of stable manure will command. The reader for whom this subject has any interest—and it is of vital importance to all who live in houses not drained into sewers, as well as to a large proportion of those who enjoy this doubtful advantage—should study this subject carefully with the aid of any one of the manuals on composting manures. To treat the subject in any detail would require the surrender of more space than can be spared in this volume. I could, moreover, add nothing of value to the mass of exact scientific information on this subject compiled by careful experimenters and accessible in many inexpensive books and pamphlets. The practical interest which this subject has for the sanitarian is this: Any substance which, left to decay in its own way, becomes a dangerous nuisance capable of exerting an influence unfavorable to health, may be rendered inodorous, and what is vastly more important, innoxious, by intimately mixing and covering it with clean, dry absorbent earth. No more trouble is required to do this than any person of refined tastes should be willing to take for the sake of decency and comfort. If the sanitary policing of a house and its surroundings is attended to from day to day, the labor will not be onerous nor exacting; and when to the benefit of more healthful conditions we add the pecuniary profit of conserving and utilizing all waste substances which can be made available for fertilizing purposes, even poverty and preoccupation cannot be accepted as valid excuses for the neglect of this important duty. *Sanitary benefits of composting.* *Profit.*

The privy next invites our attention—although it cannot usually be said to be an inviting object. This is commonly a *Privies.*

place so foul and offensive that a person not accustomed to its characteristic odor is prompted to avert his face and hold his nose when compelled to go near it. Very often the privy is set on top of the ground, with nothing to prevent its becoming a pestilent nuisance except the action of the air in drying the mass of putrefaction beneath. The soil becomes soaked by the liquid constituents of the excremental matter, and each rain may wash some of it off toward the well or spring from which drinking water is taken. The very thought is sickening, and yet the case is by no means uncommon. In every village and country town such privies are the rule rather than the exception. I have seen in a New Jersey town, in a light, porous, sandy soil, the privy located within 50 feet of the house and in close proximity to the well.

The typical country "backhouse."

Saturation of soil.

The neglected privy is a relic of barbarism which should no longer be tolerated in civilized communities. The earth closet, of which I shall speak more fully further on, should be substituted for it; but if the privy must remain, let us respect health if not decency, and compost the foulness it is built to contain. There should be no such thing as a privy vault. Under the seat there should be a box with tight joints into which everything could fall. The back of the building should be so constructed as to permit this box to be drawn out and emptied. A good shape for a box of this kind is to have the bottom slightly rounded up at one end, to which is fastened a stout iron ring so that a horse may be hooked fast to it and draw it away like a stone drag. When placed in position the bottom of the box should be covered to a depth of 3 or 4 inches with dry earth, the more absorbent the better. For greater convenience it would be well to have the seat hinged so that it can be raised, giving access to the box from the top for its entire length. With these simple and inexpensive preparations made, it is only necessary to sprinkle a little dry earth daily over the contents of the box. Properly, a quart or two should be thrown in whenever the privy is used; but this is not likely to be done unless the operation can be effected automatical-

Earth closets.

Substitute for privy vaults.

How to make an earth privy.

Disinfection and deodorization of excreta with earth.

ly, and few persons will incur the expense of providing a privy with the regular earth-closet apparatus for letting down a certain quantity of earth upon each fresh deposit of fæcal matter. I recommend this arrangement for several reasons. The most important of these are its cheapness, simplicity and efficiency. I have seen excellent results secured by placing a tight cask under each seat, with a bottom layer of earth. In connection with such an arrangement there should be a box of dry earth in one corner of the privy, and a scoop or small shovel with which to throw it in. It is some trouble to keep this box filled and to throw earth into the receptacle, but it is amply repaid. I know of nothing more disgusting to sight and smell, more nauseating to the stomach or more dangerous to health, than a typical country privy, with its quivering, reeking stalagmite of excrement under each seat, resting on a bed of filth indescribable. I feel as if it devolved upon me to ask pardon of the reader for even mentioning such a nightmare horror; but the writer upon such subjects must not stop to choose his words when attacking an evil so serious as this. Such privies as I have described are by no means exceptional. One may find them peering over the lilacs or hiding in conscious shame behind the grape arbors close beside an unfortunately large percentage of country houses occupied by people who, in all other matters, live decently and comfortably.

There are several ways of composting fæcal matter with dry earth, but I know of none better, simpler or less expensive than that I have suggested. Disinfectants may be used with advantage in connection with earth, if needed, but they are practically powerless, if used alone, to render harmless and inodorous the contents of a foul privy vault. I have tested this very thoroughly, and my conclusion is that a long-neglected privy is beyond reform by any means other than those needed to reform it out of existence. The best way to do this is to empty the vault, fill it with clean dry earth and split up the house for kindlings. I also know from experience that a summer hotel privy, used daily by a large number of people, can

Composting fæces.

Disinfectants

A sanitary privy.

be so well taken care of that it will be as free from unpleasant sights and smells as the front porch. In the case in mind a small quantity of sifted dry earth was thrown in two or three times daily by a boy, and as often as necessary the boxes were taken away and emptied in a place where their contents could be made available for further service in composting with kitchen garbage, &c. The expense was trifling and the results secured were such as to satisfy the most rigid sanitarian. The method is attended with no difficulties, and no illustrations are needed to make it plain to the simplest understanding.

Privies should not be a sole dependence. But an outdoor privy, however well kept, should not be the only convenience of its kind provided for the occupants of country houses. In dry summer weather they answer the purpose well enough, perhaps; but in wet weather, and especially in winter, their use involves an exposure which few constitutions are strong enough to bear with impunity. Women are especial sufferers from this cause; hence we find that in wet or cold weather they defer their visits to the privy until compelled by unbearable physical discomfort to brave the dangers and annoyances of a dash out of doors—for which, I may add, *Irregularity of habit and its consequences.* they very rarely wear sufficient clothing. The results of the irregularity of habit thus induced are, if possible, even worse than those attending the frequent exposures incident to greater regularity. It is not an uncommon thing for women in the country to allow themselves to become so constipated that days *Constipation induced by neglect.* and sometimes weeks will pass between stools. Physicians practicing in cities, where every provision is made for comfort and convenience, if not health, by means of indoor water-closets, tell me that irregularity in attending to the requirements of nature is a fruitful source of sickness among women. It seems to be a tendency of the sex which easily assumes the form of a habit. If this be so in cities, what can we expect in country districts, where a visit to an outdoor privy in a cold storm or when the ground is covered with snow and the air frosty is attended with a physical shock which even strong

DRAINAGE OF COUNTRY HOUSES.

men dread? Under such circumstances we can scarcely blame those women who, ignorant of the consequences to themselves, defer the performance of this important duty as long as possible. We may more justly pity them as the victims of a custom which, in this age of enlightenment, is simply disgraceful. This, however, is a subject upon which it is of little use to talk or write merely. Until we provide our families with better facilities than are now commonly enjoyed by them, the important duty of a daily evacuation of the bowels will be neglected in wet or cold weather by all who can find any excuse for so doing. *Sanitary importance of indoor commodes.*

When the need of a substitute for, or indoor supplement to, the privy is felt, the owner of a country house, if in comfortable circumstances, commonly has a water-closet put in. This obviates the difficulty of which I have last spoken, but it usually gives rise to another which, though wholly different, may exert a still wider influence for mischief. The objection to a water-closet in a country house lies in the difficulty of providing the means of effectually disposing of the matter which passes down the soil pipe. Under exceptional conditions the house can be drained into a running stream, but while this may solve the problem so far as the individual householder is concerned, it immediately acquires an interest for the community. It is possible, of course, to dispose of water-closet soil even when we have no sewer into which to run it; but this can only be done properly by separating the solid and fluid constituents of the waste, filtering the latter and mixing the former with dry earth or other material which will absorb the gases generated by its decomposition and render it innoxious. The function of water in house drainage is only that of a carrier. When it has performed its work it leaves the matter carried pretty much as it found it, and wherever the place of final deposit may be, if above ground or massed in pits, there are bred poisons which may do infinite mischief. *Water-closets in country houses. Difficulties of disposing of soil. Water only a carrier.*

A simpler, cheaper, safer and altogether more convenient because movable—apparatus than the water-closet for country *The earth closet.*

houses, is the earth closet. This device is as yet little understood or appreciated in this country. It is a machine for disposing of excreta with the least possible trouble; and so perfect is it in operation that an earth closet may remain in a bedroom or sitting room or the chamber of an invalid, and be in constant use, without making its presence known unless neglected, and without receiving other attention than an occasional refilling of the hopper with dry, sifted earth and emptying the receiver placed under the seat as often as it becomes full.

Dry earth as a disinfectant. The deodorizing and disinfecting qualities of dry earth have been known from the earliest ages. In the instructions given by Moses to the Israelites during their march through the wilderness, as recorded in Deut. xxiii, 12th and 13th verses, these qualities are recognized and put to practical use. The Chinese have also known and profited by the same facts from time immemorial. The power which dry earth possesses of absorbing the effluvia and all other noxious elements of excreta has, by the latter people, been so utilized that not only is the atmosphere in and around their dwellings kept free from contamination, but the earth itself, after being so used, becomes an excellent fertilizer, and to its extensive employment for this purpose is ascribed the wonderful and perpetual fertility of the more densely peopled regions of the Chinese Empire.

Earth closets of English origin. Earth closets are of comparatively recent origin, having been patented by the Rev. Henry Moule, an English clergyman, in 1860. Mr. Moule, who lived at a country parsonage, had been greatly troubled with the nuisance caused by the cesspool of his house, which, like many others, was situated in close proximity to the well from which the family had to draw their supply of water; and as the well was threatened with complete pollution, he made an effort to avert the danger and get rid of the nuisance. He abolished privies and water-closets, and placed small buckets beneath the seats for the reception of the excreta, the contents of which were regularly emptied into a trench made in the ground for that purpose. In a short time he made the discovery that the effect of the earth on the fæcal matter was to totally deodorize and disinfect it. This discovery led to further experiments, until he devised the mechanical means of using dry sifted earth in an ordinary closet or commode, and having patented his invention he introduced it to the public. His system has been tried with success in many places in England and in India. It has been found especially useful as applied to large public institutions, barracks, encampments, &c., and the strongest testimony has been obtained as to its complete success.

In a report to the Privy Council the following summary of the advantages of this system are given by Dr. Buchanan: *Dr. Buchanan on earth closets.*

1. The earth closet, intelligently managed, furnishes a means of disposing of excrement without nuisance, and apparently without detriment to health.

2. In communities the earth-closet system requires to be managed by the authorities of the place, and will pay at least the expenses of its management.

3. In the poorer class of houses, where supervision of closet arrangements is indispensable, the adoption of the earth system offers special advantages.

4. The earth system of excrement removal does not supersede the necessity for an independent means of removing slops, rain water and soil water.

5. The limits of application of the earth system in the future cannot be stated. In existing towns, favorably arranged for access to the closets, the system might at once be applied to populations of 10,000 persons.

6. As compared with the water-closet the earth closet has these advantages: It is cheaper in original cost, it requires less repairs, it is not injured by frost, it is not damaged by improper substances being thrown down it, and it very greatly reduces the quantity of water required by each household.

7. As regards the application of excrement to the land, the advantages of the earth system are these: The whole agricul-

tural value of the excrement is retained; the resulting manure is in a state in which it can be kept, carried about and applied to crops with facility; there is no need for restricting its use to any particular area, nor for using it at times when, agriculturally, it is worthless; and it can be applied with advantage to a very great variety, if not to all, crops and soils. After the disposal of excrement by earth, irrigation will continue to have its value as a means of extracting from the refuse water of a place whatever agricultural value it may possess, for the benefit of such crops and such places as can advantageously be subjected to the process.

8. These conclusions have no reference to the disposal of trade or manufacturing refuse, which, it is assumed, ought to be dealt with as belonging to the business in which it is produced by the people who produce it, and not to come within the province of local authorities to provide for.

From personal experience, and after the severest tests which I could devise, I can recommend the earth closet as the best, cheapest and most generally satisfactory of indoor commodes for country houses.

Price of earth closets. There are several forms of earth closets in the market. From $20 to $25 is, I believe, the price of one made after the most approved pattern, with a capacious hopper and an arrangement for discharging a fixed quantity of earth into the receiver. Those who are able and willing to pay this price will get a good *Home-made earth commodes.* article, with full directions for its use and care. For the benefit of those who are not, I will say that a convenient earth closet can easily be made, at small expense and without infringing anybody's patents, by any person with intelligence enough to build a hen-coop. My own experience in building and managing an earth closet may not be without interest. I made it of pine boards in the shape shown in Fig. 26. It was simply a box with two covers and no bottom. The under cover, which served as the seat, was hinged to the edge of the box, and the upper cover was hinged to the lower, so that they could be

DRAINAGE OF COUNTRY HOUSES. 273

raised singly or together, as desired, without interfering with each other. Under the seat, and standing upon the floor, I

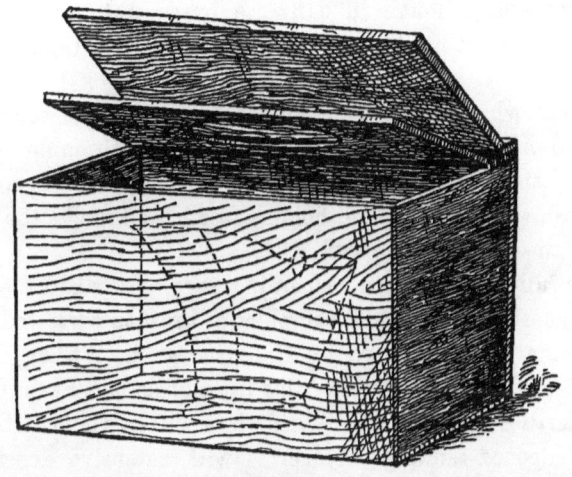

A cheap and convenient form of earth commode.

Fig. 26.

placed a galvanized iron coal-hod. A tin pail, full of dry, sifted earth, stood beside it. When two or three inches of earth had been sprinkled upon the bottom of the coal-hod the earth closet

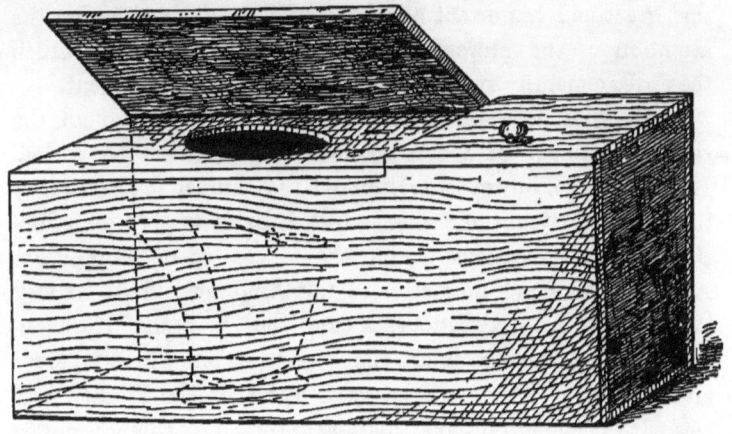

Fig. 27.

was ready for use. The whole cost of the apparatus, including a large coal-hod, did not exceed $3·50, but it was as satisfactory
18

as one could be. A small shovelful of earth was thrown in when the closet was used, and it was perfectly free from unpleasant odor, though in daily use by several persons. The only attention it needed or received was to empty the hod when full.

A somewhat more convenient shape for the box would have been to make it long enough to admit of partitioning off one end for an earth reservoir, as shown in Fig. 27. This would dispense with the pail for holding earth and make the whole apparatus complete in itself.

For fuller information concerning such closets and their use, the reader is referred to a pamphlet of great interest and value written a few years ago by Col. George E. Waring, Jr., of Newport, R. I. The title of this little book is "Earth Closets and Earth Sewage." Mr. Waring is a writer who combines a knowledge of sanitary engineering with extensive experience, a habit of careful and intelligent observation, and a literary style so pleasant that even the casual reader is interested and instructed. If a copy of his pamphlet on earth closets were placed in the hands of every country physician, I am satisfied that great and important benefits would result in drawing the attention of the profession to many things concerning which they are, generally speaking, either ignorant or indifferent.

Geo. E. Waring, Jr.

If there is no water-closet to complicate the problem, the sanitary drainage of a country house is somewhat simplified. It is a mistake, however, to suppose that human excrement is the only constituent of sewage which is liable to give off offensive and poisonous gases during the process of decomposition. A sewer into which no matter of this kind ever finds its way is, under ordinary conditions, as dangerous to health, if unventilated except through house connections, as one which receives all the waste of a town. The waste water of the kitchen carries with it enough organic matter to breed pestilence under favorable conditions, and for this reason the proper drainage of a country house which has a sink in the kitchen is a matter of prime importance as affecting the health of the inmates.

Sanitary drainage of country houses.

Organic matter in waste water.

DRAINAGE OF COUNTRY HOUSES.

At the back doors of farm and village houses we commonly find a serious evil, either in a defective drain or in the absence of any drain at all. In the latter case the "slops" are commonly thrown out upon the ground and left to take care of themselves. The ground, instead of being soft and absorbent, is bare, hard and often covered with mould. To a person unaccustomed to it the smell is nauseating. If a drain is used it generally ends nowhere, and is often not more than 10 or 12 feet long—a little pool at the end catching what passes through it. The miscellaneous refuse of the kitchen finds its way into it and must go through the usual process of decay in the drain or about its mouth. When we find such a slovenly method of disposing of the kitchen refuse, we may take it for granted that wash water from the bedrooms is thrown out of the window and chamber lye poured on the grass.

Back-door nuisances.

Drains.

The common method of draining country houses of the better class in the United States is into stone or brick cesspools. The same system is employed in a majority of villages and unsewered towns when any provision is made for house drainage. As the rule, such cesspools are merely unventilated cisterns with bottoms and sides more or less porous, through which a part of the foul water discharged into them escapes to saturate the surrounding soil. That leaching cesspools are wholly bad is a statement which I can make without fear of intelligent contradiction. Such cesspools are a fruitful source of disease and death in rural neighborhoods where they have been introduced. Sewers are bad enough, even under the most favorable conditions, though for the present they seem to be necessary evils in cities and populous districts; but leaching cesspools at their best are liable to be worse than sewers at their worst, since they are not channels to carry away filth, but receptacles for its storage, wherein we can manufacture our own supplies of sewer gas and conduct it into our houses through the waste pipes which we fondly imagine are effectually sealed against it by water in the traps. How much of a

Leaching cesspools worse than sewers.

dependence this is has already been explained in a previous chapter. Dr. Playfair, in an excellent address before the British Social Science Association, speaks of such cesspools as follows: "Instead of allowing garbage to be freely oxidized, or applying it to plant life, which is its natural destination, we dig holes close to our own doors and cherish the foul matter in cesspools under conditions in which air cannot enter freely, and therefore the most favorable to injurious putrefaction. We forget the superstition of our forefathers, that every cesspool has its own particular evil spirit residing within it, and we are surprised when the demon emerges, especially at night, and strikes down our loved ones with typhoid fever or other form of pestilence."

I am not prepared to advocate the abolition of the cesspool, as it is still indispensably necessary under a great variety of conditions; but in every case it should be made as tight as a bottle to start with. Any mason can build such a cesspool, and the method and materials to be used need not be described. The end to be secured is to prevent leaking, and there is no more trouble in attaining this in a cesspool than in a cistern. It would be well to make it so small that it should need to be emptied every few weeks, and provision for such emptying should be made by means of a suitable pump always ready for use. There are many in the market which will do this work admirably. The cesspool should be dug as far from the house as convenient—say 100 feet—and the top should be left open so as to afford a free vent for gases which must otherwise work their way back through the pipes into the house. The connection with the house may be made with glazed tile— preferably Scotch—with the best cement joints. There should not, in my judgment, be any traps except those in the branch waste pipes inside the house, and, as in city houses, the main waste should be ventilated above the roof. The cesspool may be covered in whatever way is most convenient or ornamental, provided an abundant vent is left, as before specified. My own plan, in venting cesspools near houses, has been to cover the top with a flagstone, in which a hole is cut about 10 inches square. Into this I set a wooden box or chimney about 6 feet high, with a door in one side, which, when opened, discloses a series of alternating shelves extending half way across the chim-

Ventilation for cesspool.

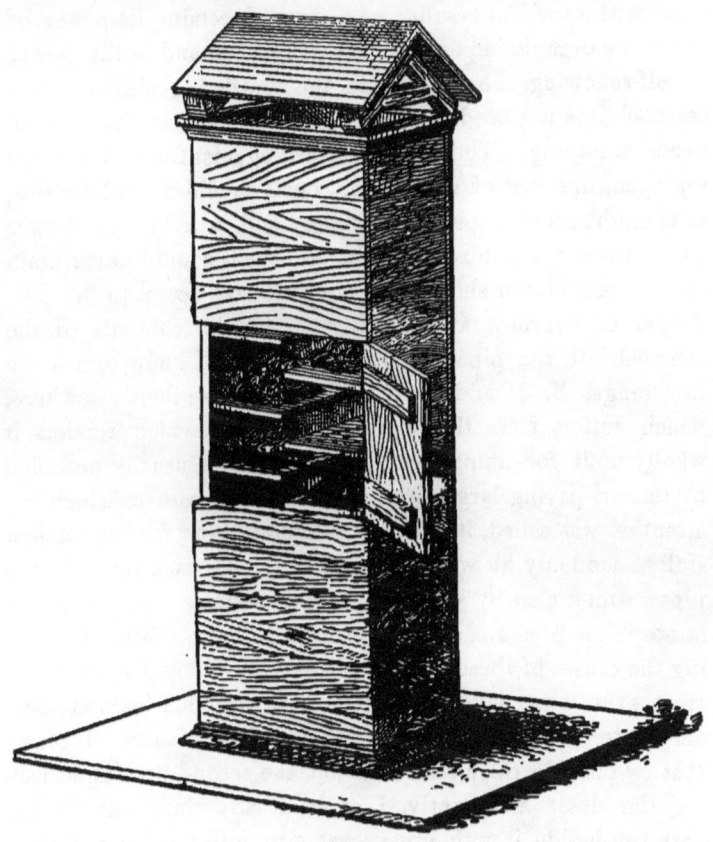

Fig. 28.

ney, as shown in the drawing, Fig. 28. The top of the chimney is finished with a cap of metal or wood, merely to exclude snow and sleet. Upon the shelves charcoal is placed, the door is shut and fastened with a hook, and the arrangements are complete. The charcoal should be renewed occasionally. It absorbs all the offensive and hurtful impurities in the gaseous exhalations from the cesspool, and without interfering with the ventilation keeps the thing from becoming a nuisance. This is similar in principle to the ventilating shaft recommended by Mr. Latham, and applied by him to sewers in several English cities with excellent results. As charcoal retains its power of absorbing organic impurities for a long time, and as this power is self-renewing, the charcoal in the ventilating chimney of a cesspool does not need to be changed oftener than the cesspool needs emptying. The next best arrangement is to leave the top open, like that of a curb well, but I should not advise this, as it might prove a source of anxiety and, possibly, of danger where there are children. Roofs and sheds should never drain into a cesspool, nor should storm water have access to it. The danger of overflow and backing up of the contents of the cesspool into the pipes should be avoided. I know of a house in Orange, N. J., and have heard of many others elsewhere, which suffers from this evil to an extent which renders it wholly unfit for human habitation, although usually occupied by tenants paying large rentals. In the one case to which my attention was called, it is no uncommon thing for the kitchen sink to suddenly fill with dirty water forced back through the pipes, sometimes by overflow and sometimes apparently by atmospheric pressure. I did not have an opportunity of studying the causes of these phenomena as carefully as I wished, but from even a partial examination of the system I learned some very interesting facts which greatly surprised me. I found that by opening the closet valve on the second floor and flushing the closet abundantly, I could at any time half fill the bath tub beside it with greasy water of milky color, evidently from the cesspool. As this subsided, water of similar color and smell would rise a few inches in the basins and sink on the first floor. The tenant of the house, who was unwilling to incur any expense in the matter, said that emptying the cesspool only helped the matter for a few days, and that it

DRAINAGE OF COUNTRY HOUSES. 279

recurred again long before the cesspool was refilled. As the cesspool was sealed tight and the pipe system was wholly unventilated, it was not difficult to account for the phenomena to my own satisfaction.

A very common method of drainage in some parts of the country consists in discharging the main waste into a blind ditch filled with cobble stones and covered with earth. This is open to the same objections as the leaching cesspool system. These ditches do not commonly fill up with water, but the interstices between the stones become choked with grease and solid filth, and long before the outflow of the house waste is checked the gaseous products of organic decomposition, formed in great volume, work back through the pipes and past their seals into the living and sleeping rooms of the house. So far as healthfulness is concerned, I would as soon carry a speaking tube from my bedroom into a grave where some body lay rotting as carry an unventilated waste pipe from my wash basin into such a subterranean grave of decomposing filth as this. *Blind drains.*

Col. Waring, from whose writings I have before had occasion to quote in this chapter, employs at Ogden Farm, Newport, R. I., and has introduced in Lenox, Mass., and elsewhere, a system of drainage which will be found to answer admirably under favorable conditions. It may be described as follows: The house drainage is discharged into a Field flush tank, which will be described hereafter. This tank, when filled, empties itself into a system of drain tiles laid with open joints, consisting of one main 50 feet long and ten lateral drains 6 feet apart and each about 20 feet long. These pipes are laid from 10 to 12 inches below the surface. In describing the details and practical workings of this system, Col. Waring says: "My suggestion is to use this system usually when there is no public water supply, with an area of 2500 square feet for each household. If there are ten persons in the family, there will be an area of 250 square feet *The intermittent down ward filtration system.*

for each. The character of the soil will have much to do with the process of purification, but probably not much with its efficiency. Heavy clay soils exert in themselves a stronger absorptive action than do porous soils, but porous soils are much more open to the admission of air, with its destructive oxygen. Naturally, the system will work best when the ground is not frozen, and during the season of vegetation we have the further advantage, which I believe to be only a sec-

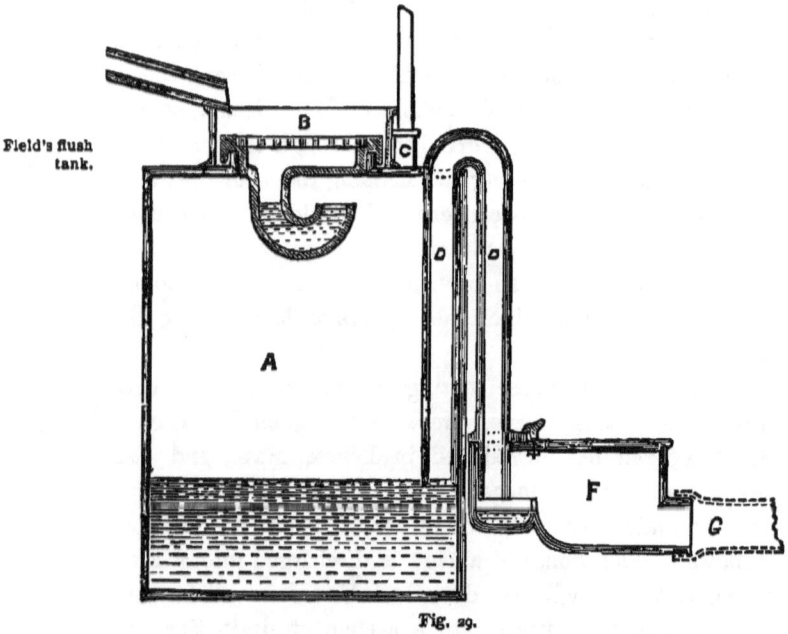

Field's flush tank.

Fig. 29.

ondary one, that the products of decomposition are taken up by the roots of plants. If not so taken up they will be entirely dissipated. Let us suppose that a household of ten persons use 300 gallons of water per day. This will give 3 gallons of sewage to 25 square feet of ground. If there is no gravel streak to lead the descending water of heavy rains to the well, until we reach a depth of only 6 feet we shall have 150 cubic feet of earth to filter 3 gallons of water per day."

DRAINAGE OF COUNTRY HOUSES.

The flush tank, which forms an important feature of this system, is a self-emptying cesspool of small size, shown in sufficient detail in the sectional drawing marked Fig. 29.

A cylindrical tank, A, has an opening in the top with a movable cover and grating, B, whereby access for cleansing purposes is given to the inside of the tank, but also acting as a trapped inlet for the flow from the sink pipe, which discharges over the grating, the tank being placed outside the house, so that direct communication between the drains and the interior is interrupted, completely preventing the entry of foul air. The top of the tank is also provided with a ventilating pipe, C, and a syphon pipe, D, the outer and lower extremity dipping into a discharging trough, F, which consists of a small chamber fixed so that it can be turned round, with the object of setting its mouth in any direction that may be requisite to connect it with the outlet pipes G. A movable cover provides for access to the mouth of the syphon when requisite. The position of the ventilating pipe C may obviously be varied according to convenience. *Construction and operation*

The trapping of the inlet and the discharging power of the outlet, are the two chief features of this apparatus which merit attention. As regards the former, it is to be noted that some considerable difficulty was at first experienced, owing to the inlet trap becoming emptied inductively by the suction due to the rapid discharge of water from the tank whenever the syphon came into action. By means of the arrangement of trap and air pipe shown in the engraving, this difficulty has been entirely overcome. The bend of the trap being located below the top of the tank, the suction, which could only arise when the level of the surface of the water had sunk as low as the top of the bend, is prevented by the supply of air from the air pipe. It will be observed that the inlet is doubly trapped, being water sealed by the flanges on the rim in addition to the bend of the trap. *Features of merit.*

As regards the action of the syphon, it may be remarked that to bring an ordinary syphon into action it would be necessary *Action of the syphon.*

that sufficient water should be run in rapidly to raise the level of the surface in the tank above the top of the syphon bend, so as to expel all the air quickly, which would require a considerable volume of water; and, on the other hand, in the case of small dribblets of water flowing in at intervals when the tank is full, they would simply trickle away over the lip of the syphon, and so the tank would remain full and the syphon continue in-operative. It is by the peculiar construction of the discharging trough, which is a special and important feature of the tank, that aid is given to assist small quantities of liquids in bringing the syphon into action, instead of dribbling over the syphon without charging it, as would otherwise happen, as explained; and this is attained by checking the efflux of fluid from the syphon outlet by the agency of a peculiar arrangement of weirs at the discharging orifice or mouth, thus obviating all the difficulties. The mouth of the syphon pipe dips into the discharging trough to the level of the top of the weir, and the weir itself is provided with a notch, the object of which is to prevent the partial or false action of the syphon, such as would result if its mouth were entirely sealed when the flow of water is not adequate to fully charge the syphon. For such a case the notch is so proportioned—as determined practically by experiment—that very small quantities of water, which are insufficient to fully charge the syphon, may run away through the notch without sealing the mouth of the syphon; whereas, on the other hand, an adequate charge, being more than will pass freely through the notch, accumulates behind the weir, sealing the syphon so as to generate its full action and initiate a complete discharge. So effectual is that action that a mere hand-bowlful of water or slops thrown down a sink and flowing into the tank when full, suffices to set the syphon in operation. This device for securing an intermittent automatic syphon action is singularly simple and effective. So soon as established, the contents of the tank are completely discharged with considerable flushing force, producing an efficient scour in the outlet drains.

As an idea of the operation of this system can best be had from a description of its practical working, I further quote from Col. Waring, as that gentleman is entitled to the credit of introducing the system into this country and making the profession acquainted with its advantages. In an account of the drainage of his own house, he says:

"Seven years ago last October, when I built my present house, I applied this method there in the most thorough way, and have been watching it with great care with a view to what I might learn from it from that time to this. I do not hesitate to pronounce it absolutely perfect. I am satisfied that it affords relief which is open to every one who has even a little bit of ground adjoining his house. I would say, by the bye, that I have no water-closets in the establishment; we use earth closets only; so that my experiment has not been complicated by that element. At the same time there is no practical difficulty; there is no reason why that may not be taken care of as well as the other.

"Outside of my kitchen the waste pipe of the kitchen sink discharges into a flush tank—that is, a vessel holding about a barrel of water supplied with a syphon which comes into action automatically; it holds back all the flow of the kitchen sink until it becomes entirely full; then almost instantly—within three or four minutes—it discharges the whole of that volume, which in my case is about a barrel of water, rapidly into the drain and drives or carries everything forward with it. The water from the baths and the housemaid's sink and other things enter the drain further down. If they do deposit any small amount of matter, this flow, which occurs as often as two or three times a week, is sure to carry everything forward. This goes to a settling basin, which is very small, having a capacity only of about 40 or 50 gallons, and which is simply for the purpose of restraining the grease which floats on the surface of the water and the solid matters which settle at the bottom. The overflow from the settling basin is through a pipe which points down below the surface, so that whatever enters this pipe must enter it below the scum and above the deposit, and whatever is discharged from this settling basin is liquid, and that liquid is carried forward through a tight pipe a distance of about 40 feet from my library window, and there it turns and runs parallel with the house for a distance of 60 feet. At intervals of 6 feet, leading from that like a gridiron, are drains of ordinary agricultural tiles; these drains which lead from it are ten in number; they are 20 feet long, loosely meeting together at the ends with no cement; they lie 12 or 13 inches below the surface of the ground, which is, I am satisfied, somewhat too deep—9 or 10 inches would be better. Whenever that flush tank discharges, it flows into a settling basin and displaces an equal quantity of liquid matter from there, which is at once driven forward and is sufficient to gorge these tiles from end to end; the contents instantly begin oozing out at the joints, and the overflow in a very short time is dispersed into the ground. The water of course settles, for this must be on tolerably drained land; it would not do to try this on the surface of a swamp which is saturated below. The water settles through the soil, thus finding an outlet, and the soil through which it passes filters out the foul matters. Immediately the water passes away fresh air enters from the surface; and by the well-known concentrated oxidizing power of porous matters, whether powdered charcoal, earth or whatever it may be, an entire decomposition is effected of this foreign matter, so much so that after five years, there being from defective work an occasion to take up a part of this system of drainage, I took up the whole and gave it a thorough examination, and in no place could you detect in the earth which lay adjacent to these tiles, in which they were immediately encompassed, either by appearance or odor, the slightest difference from ordinary fresh-smelling garden mould. This has been going on, as I say, since seven years ago last autumn, for a household of six persons, with rather a copious use of water, and there has been no other means adopted.

Depth below surface.

Filtration through the ground.

"I should not, of course, on my own single experiment, venture to recommend this, as I have done frequently, to the public as being worthy of adoption. Its use has extended very much. I applied it last year to the sewage of the whole village of Lenox, in Massachusetts; and in England it is being adopted for the sewage of country houses far and wide, and is based on the principle which is thought by many English engineers to promise the only relief that they can have from their sewage. When I am describing this, the question which is almost universally asked is, What becomes of the solid matter and grease in the settling basin? At first I used to have it taken out and buried about once in three months—dug a trench in the ground near by, cleaned out the settling basin and buried its contents in the trench. But once, only a week after cleaning it out, I had occasion to empty it again for another purpose and found that it was as foul as it had been after a longer interval. That was about three years ago. Since that time the settling basin has never been opened except for inspection, and its condition remains always the same. The explanation is perfectly simple: The solid matter at the bottom of the tank is decomposable matter and is constantly passing itself off in solution in the water that flows away; and the matters which are decomposing are very strong producers of ammonia, which acts upon the under side of the floor of grease and converts that into soap, which in its time passes off." *Town sewage works. The settling basin.*

Having had three years' experience with this system, so far as its essential details are concerned, in draining my own house, I have no hesitation in expressing the opinion that under favorable conditions it will work satisfactorily and be found an improvement on any other system which can be contained within the restricted limits of a village lot or villa site. There seems to be no reason why it should not work equally well on a larger scale, and in the case of Lenox I am informed that it does. English testimony is also strongly in its favor, and nowhere else has it been tested with equal thoroughness nor under so great *The author's experience.*

a variety of conditions. When the conditions are unfavorable or householders are unwilling to venture even so simple an experiment in sanitary engineering, I should recommend the tight, well-vented cesspool already described.

When there is no plumbing work in a house and no facilities are needed except those which afford a safe and convenient means of disposing of dish water and kitchen slops, a cheap and simple device is a box filled with absorbent earth. The

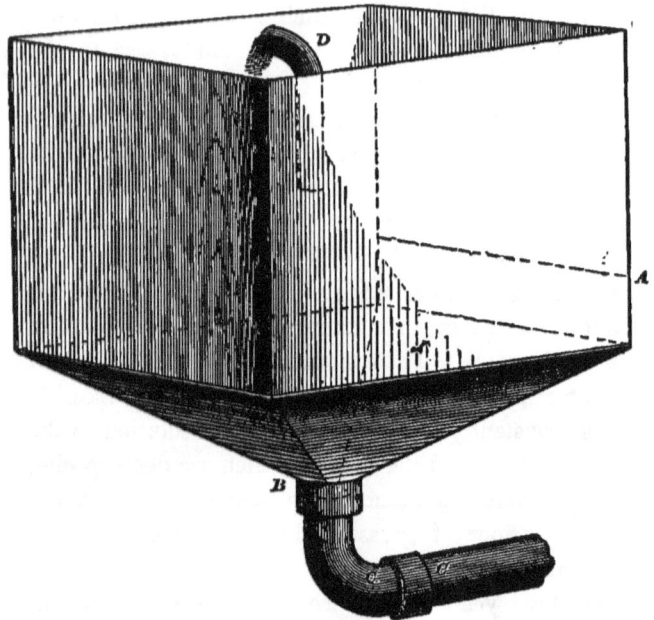

Fig. 30.

function of this filtering tank is to remove from the waste water all matters which can readily be strained out and retain them in such shape as to admit of their subsequent utilization for fertilizing purposes. I have generally found a tank 4 feet square amply capacious. This gives us a cubic contents of 64 square feet. The shape of the tank is not a matter of great importance, provided the bottom is so inclined that all the water flowing into it shall find its way to the point at which an outlet is provided. The shape I prefer is shown in Fig. 30, and when one is made especially for this use it might as well be of this form as any other. As will be seen, the bottom has the shape of an inverted pyramid, formed of four triangular pieces joining the straight sides. For convenience in emptying, one of the sides is made in two parts, united with hinges at the line A. The tank may be set into the ground to the line A, which has the advantage of bringing the pipe C, which carries off the filtered water, below the depth to which frost will penetrate the ground except in high northern localities or exceptionally cold winters without snow. At the point of discharge, B, which is formed by nailing a collar of zinc to the bottom pieces, it is well to have some kind of strainer—either a perforated metal sheet, a piece of wire cloth, a block of soft peat or anything that will serve the purpose. A sabot of hay or straw will answer as well as anything else. In the bottom of the tank place a quantity of broken straw, packing it tightly. On this throw about 12 inches of dry, absorbent earth, leaf mould or any kind of soil other than sand or clay. The tank is then ready to receive so much of the drainage of the house as requires such treatment. The pipe D is the waste from the kitchen sink. The outflow from this pipe is water more or less charged with organic matter. The water passes through the filtering material in the tank, but the grease, the bits of meat, the vegetable particles and all the solid or semi-solid constituents of the outflow are retained in the tank. As often as may be necessary to prevent the escape of any offensive odor, or even the characteristic greasy smell which is usually noticeable in a kitchen drain, throw in more earth until the tank is full. In warm weather a little powdered gypsum (ground plaster) thrown in would be found useful in absorbing and retaining the ammonia formed from the decomposition of substances containing nitrogen. The matters which come out of a kitchen through the sink waste are not in themselves offensive or injurious. They only become so by decomposition, a process which does not commonly begin under 24 to 48 hours. If kept covered with any dry, absorbent material, the products of decomposition are all retained, and this will continue until the earth in the tank is saturated and ceases to absorb and deodorize the matter which it holds, and then it will require renewing. How often it will need this depends upon circumstances. In the instances I have in mind in which this system has given perfectly satisfactory results, the tanks have not required to be emptied and refilled oftener than once in two months. In one case four times a year was often enough. A layer of earth an inch thick is commonly sufficient to throw in at one time, and this may be done two or three times a week in very warm weather when decomposition goes on rapidly. With good management less attention than this will keep the tank in good condition. The pipe C is common earthen tile. A few feet away from the tank and for the rest of its length it may be laid with open joints. It may extend under a garden or lawn, or indeed anywhere. Water which flows into it leaks out at the joints or through the pores of the pipe, is absorbed by the soil and appropriated by vegetation.

<small>Renewal of contents of tank.</small>

<small>The discharge pipe for filtered water.</small>

Such a tank as I have described may be placed anywhere. I have never found any objection to placing it against the foundation of the house, so that the kitchen sink waste needed to be only a few feet in length. To exclude rain it requires some kind of light spreading cover, raised high enough above the sides to give the fullest ventilation and permit free evaporation. When full, one side is lowered and the contents are shoveled into a wheelbarrow and removed to the compost heap.

<small>Position of tank.</small>

<small>Cover.</small>

Experience has shown it to be a great mistake to suppose that isolation, with good surroundings and plenty of fresh air out of doors, will insure good health. The fallacy of this notion was forced upon my attention during the summer of 1876. In the course of a ramble among the hills of Orange county, N. Y., I made the acquaintance of a farmer and his family who had lately experienced a terrible visitation of typhoid fever. The farm occupied one of the most beautiful locations I had ever seen. It lay upon a mountain top about 1440 feet above the

<small>Country districts not always healthy.</small>

<small>Typhoid fever on a mountain top.</small>

sea and some 600 feet above the lake in the valley. The view was extensive in all directions; the formation was such as to insure the best natural drainage, and the winds from every quarter of the heavens had full sweep over the whole place. Any one would have fixed upon this spot as a sanitary paradise, but the fever had found it, though there was not another dwelling from which the seeds of contagion could by any chance have been carried. The cause was not difficult to find, however. My friend the farmer made dairying his principal business. In the barn, almost immediately adjoining the house, he wintered his cattle, and as it was easier to carry water for the family than for the stock, he had sunk his well in the barnyard. Gradually but surely the sources of his water supply had been poisoned, and with all natural conditions favorable to health in a peculiar degree, he had in his ignorance brought upon himself worse evils than he would probably have encountered in the dirtiest neighborhood of the bad-smelling city. His folly had cost him a son stricken down in the full vigor of early manhood, and left himself broken in constitution and forever enfeebled by the scourge which had brought him and still others of his family to death's door.

People who leave convenient and comfortable city houses to seek pure air and healthful surroundings in the country, often find that they have made a sad mistake. Fresh air and sunlight are certainly great purifiers and disinfectors, but these potent agencies will not purify our wells nor arrest decomposition. The absolute dependence often placed upon them in country districts accounts for the fact that these districts, when fairly well settled, are rarely free from epidemic diseases of one kind or another. It may seem like an extravagant statement, but I have no hesitation in saying that were hygienic precautions as generally neglected in New York city as in the country districts, the columns of our daily newspapers at any season of the year would read like chapters from De Foe's history of the London plague. *Seeking health in the country.* *Neglect of sanitary precautions in country districts.*

No excuse for epidemics. With every possible facility for keeping in good health, the residents of country districts and of our smaller towns have no excuse for the fevers and other epidemic diseases which are so common nowadays, and which are so very generally caused by willful violations of hygienic laws. We can hardly imagine any reason why a farmer or the tenant of an isolated house in the country should ever see a case of typhoid. Yet typhoid fever is common, and people wonder at the Providence which robs them of their friends and children, and try to be resigned to the decrees of a higher power, when the cause is really to be found in their own disregard of the laws of health—laws of which no one has any right to be wholly ignorant in this age of enlightenment.

CHAPTER XI.

Water Supply in Country Districts.

In selecting a place of residence or a building site in the country, care should be taken to secure an abundant supply of pure water. Other things are important and should be given due weight, but the quality of the water supply upon which dependence must be placed, is of prime importance. It is possible to correct nearly all the evils prejudicial to health if the proper means are used, but we cannot find good water where it does not exist. To the neglect of this precaution we may, I think, attribute much of the sickness in populous suburban districts surrounding our principal cities. No country house is desirable as a place of residence which does not at least have a good well or an unfailing spring of pure water near it, and in such a position that it is free from all danger of contamination by surface water flowing into it, or by impurities reaching its sources through permeable strata of the soil. There is a great deal of valuable literature on the subject of water which is easily obtainable by those who seek it. In a work of this kind it is necessary to discuss the subject somewhat briefly. *Importance of good water in country places.* *Bad water a cause of sickness.*

In judging of the quality of water and its fitness for drinking and general domestic use, the principal points to be determined are: *Determining the quality of water.*

1. Total solid residue.
2. Hardness, temporary and permanent.
3. Chlorine.
4. Nitrogen in nitrates and nitrites.
5. Ammonia and organic matter.
6. Metals.

Characteristics.

The first is easily determined by evaporating a given quantity, say 3 ounces, on a water bath to dryness, and drying the *Solid residue.*

residue in an air bath at 266° Fahr. The total operation, according to Wanklyn and Chapman, requires about an hour and a quarter.

Hardness. By the hardness of water is meant the quantity of soap it will consume in forming a lather. Although hard water is not usually considered unwholesome, it is quite unsuited for general domestic use. The quality of hardness results chiefly *Determination of hardness.* from the presence of lime and magnesia in the water. There are several methods of determining hardness, the simplest of which consists in preparing an alcoholic soap solution, of which a given quantity is just capable of neutralizing one gram of carbonate of lime. For the benefit of those who have access only to Troy weights and apothecaries' fluid measures, I give the following simple directions for determining the hardness of water with United States weights and measures. Weigh out 8·88 grains pure fused chloride of calcium, made by dissolving calc spar in pure hydrochloric acid, and dissolve it in 32 fluid ounces of distilled water. Next prepare a solution of hard white soap in strong alcohol, filter and add an equal volume of water. Take 4 fluid drachms of this solution, and place in a bottle with 4 fluid ounces of distilled water. The standard solution of chloride of calcium is added to this from a graduated pipette until on shaking the frothing stops. If more than 4 drachms of the standard solution of chloride of calcium is required, dilute the soap solution with just enough 40 per cent. alcohol to make 4 drachms of the lime solution neutralize 4 drachms of soap solution in the presence of 4 ounces of water. In determining the hardness of a natural water, measure out 4 ounces of the water and place in a glass stoppered bottle, then add sufficient soap solution to produce on shaking a permanent lather. The number of drachms of soap solution consumed will indicate the number of grains of carbonate of lime (or its equivalent of magnesia), in a United States gallon of 231 cubic inches, allowing 128 fluid ounces, or 1024 drachms, to the gallon.

Permanent hardness. The determination of the permanent hardness is made as above after boiling the water for an hour, distilled water being added to replace that which evaporates. The difference between total and permanent hardness is equal to the temporary hardness.

Chlorine is usually determined volumetrically, by means of a standard solution of nitrate of silver, and requires some skill and experience to insure accuracy. If 7·37 grains nitrate of silver be dissolved in 33 fluid ounces distilled water, 0·033 ounces of the solution will precipitate 0·0015 grains of chlorine. About 0·0077 grains of neutral chromate of potash in solution is dissolved in a measured quantity of the water to be tested, and the standard silver solution added, drop by drop, until a permanent red color is noticeable. Water which contains a large amount of chlorine may have derived part of it from sewage. *Chlorine.*

Nitrates and nitrites are not easily determined, and no method now in use would prove of any value to a person not a skilled chemist. The presence of even small quantities of nitrites is objectionable, and for these the following qualitative test will suffice. A very thin paste is made of starch, and to this is added a little solution of iodide of potassium. Iodide of potassium is a solid, and the solution may be prepared by taking a very small piece, say half as large as a pea, and dissolving it in a test tube two-thirds full of water. A gill of the water to be tested is taken, and about five drops of dilute sulphuric acid are added. The acid is previously diluted by mixing a little of it with an equal weight of water in a beaker. A little of the water to be tested, thus acidified, is poured into the mixture of starch paste and iodide of potassium. If it turns blue, a nitrite is present. Otherwise there are no nitrites in the water. *Nitrates and nitrites. Test for nitrates.*

Nitrates are thus detected: to a quarter of a gill of the suspected water in a small beaker, add half a gill of pure concentrated sulphuric acid, and warm the mixture on a sand bath until the temperature is 140° Fahr. While the mixture is still warm add a few drops of an extremely dilute indigo solution. If the color of the indigo disappears immediately, even on *Tests for nitrites.*

repeated addition, a nitrate is present. The indigo may be purchased in solution, and should be ordered as sulphate of indigo solution. A very little of the liquid added to a pint of water gives the dilute solution, which should be used as above.

Ammonia and organic matter. Ammonia and organic matter are especially objectionable in potable water, and their quantitative determination is as important as it is difficult. The Nessler test is a very delicate one for ammonia, as it will detect one part of ammonia in 20,000,000 parts of water. Ammonia may be concentrated by distillation, for if 67·6 fluid ounces of water be distilled, nearly all the ammonia contained in it will pass into the first 3·4 ounces of distillate, thus rendering the test ten times more delicate than before. The Nessler reagent is made by dissolving 77·19 grains iodide of potassium in a small quantity of hot distilled water, and adding to it, while on a water bath, a solution of corrosive sublimate until the red precipitate no longer dissolves; filter and add 231·57 grains of solid caustic soda or 308·76 solid potash dissolved in water; dilute to 33·8 ounces, and add 0·169 ounces of saturated solution of corrosive sublimate; allow to subside and decant the clear liquid. It gives a brown color with ammonia.

The Nessler test.

Quantitative determination by Nessler's method. In using the Nessler test for quantitative determination, 0·05 ounce of the reagent is added to 3·4 ounces of the water to be examined, and the color observed. The same quantity of reagent is added to a given quantity of a standard solution of ammonia also diluted to 3·4 ounces, and the colors compared. The standard ammonia solution is made by dissolving 0·6 grain sulphate of ammonia, or 0·49 grain chloride of ammonium in 33·8 ounces of water.

The presence of any considerable quantity of ammonia is almost certain proof of sewage contamination, as urea is readily convertible into carbonate of ammonia.

Test for organic matter. One of the simplest tests for organic matter in water is made with permanganate of potash. This salt is remarkable for its great coloring power, a very small quantity producing a purplish red color in a great quantity of water. To perform the test, about 3 grains of the salt should be dissolved in one pint of pure water. A very pure water can be obtained by melting ice. About one pint of the water to be examined, *not concentrated*, should be introduced into a colorless glass beaker, and 5 drachms of dilute sulphuric acid added. The acid should be so diluted as to consist of one part acid and five parts water, by measure. The permanganate should then be added, a few drops at a time. If organic matter is present, the colored solution of the permanganate of potash will be decolorized as soon as it touches the water, and a brown deposit will slowly settle to the bottom. By continuing to add the colored solution until it is no longer decolorized, we can form some idea of the amount of organic matter in the water.

If water examined for organic matter be allowed to stand for a day, a sediment will often settle to the bottom. If a portion of this sediment be dried and burnt in a porcelain dish over a spirit lamp, it will smell bad if there is animal organic matter present. It will turn black if there is organic matter of any kind present. This of course shows organic matter in a state of suspension in the water. The permanganate test must be resorted to to discover organic matter in solution. *Burning the sediment of water.*

A very simple test for organic matter in water consists in dissolving therein a small quantity of white sugar. In a few days if sewage, urine, albumen or any other organic impurity be present, the water becomes white and milky from the development of certain fungoid growths. Prof. Frankland, F. R. S., presented a paper on this subject to the Chemical Society of London, as long ago as Feb. 2, 1871. In this paper he advanced the theory that these germs are everywhere present in the atmosphere, but that they cannot develop in the sugar solution without the presence of phosphoric acid or some compound of phosphorus. The following comparative tests were recently made in this city to determine the value of Prof. Frankland's suggestion in a practical way: Four 8-ounce bottles were filled *The sugar test on organic matter*

Experiments with Croton sediment. with water, and to each was added 15·44 grains of powdered sugar. The first of these solutions, which contained boiling distilled water and no air, remained unchanged during the whole experiment, lasting 50 days. No. 2, which contained Croton water and a little air, began to exhibit a white sediment in nine days, which seemed to adhere to the bottom of the bottle. In two days the third solution, to which had been added 0·17 ounces urine (or about 2 per cent.), had a milky look; in three days a heavy froth on top, and in eleven days was perfectly opaque and contained small white flakes; at the end of a month still heavier deposit and cloudy; in 50 days less opaque, heavy sediment. To the fourth solution was added 0·17 ounces of a solution of phosphate of soda. The changes were more marked than in Croton alone, but far less than in that containing urine, showing that some other ingredient than phosphorus aids in producing the change. Neither 2 nor 4 were opaque at the end of 50 days unless shaken to diffuse the heavy sediments in them.

A trace of organic matter not necessarily dangerous. A trace of organic matter does not necessarily disqualify water for general use. The following table shows the number of grains of organic matter in the water supplied to some of our American cities. The insignificance of the quantity will better appear if we bear in mind the fact that a gallon contains 70,000 grains of water:

Percentage of organic matter in waters supplied to cities.

New York (Croton)..........................1·97
Brooklyn (Ridgwood).........................1·43
Boston (Cochituate)..........................1·22
Philadelphia (Fairmount, Schuylkill)...........2·06
Albany (from the hydrant)....................3·96
Troy   "   "   ....2·30
Utica   "   "   ....1·64
Syracuse (New Reservoir)....................3·08
Cleveland (Lake Erie).........................2·62
Chicago (Lake Michigan)......................1·81
Rochester (Genesee River)....................2·12

Schenectady (State street well)..................4·00
Newark, Jersey City, Hoboken, Hudson City (Passaic River)................................4·90
Trenton (Delaware River)....................0·95

Of the metals, lead and zinc are the most dangerous and the most frequent. As directions for the determination of the various metallic salts are given in a previous chapter, they need not be repeated here. Metals.

Some natural waters contain sulphuretted hydrogen acid, which is easily detected by smell, and by its action on bright silver coin or on paper dipped in acetate of lead solution. Sulphuretted hydrogen.

Carbonic acid may be present either in a free state or combined with lime and magnesia, for bicarbonates. In either case it is expelled by boiling. To ascertain whether any of the carbonic acid is in a free state a strip of turmeric paper is employed, together with a freshly prepared solution of clear lime water. If the addition of a single drop of lime water to the water to be tested causes it to turn the turmeric paper brown, no free carbonic acid is present. If it is necessary to add several drops of the lime water before this action takes place, the quantity of free acid is quite large. Dr. Von Pettenkofer also employs for this purpose rosolic acid. Free carbonic acid is seldom if ever present in the waters of limestone regions. Carbonic acid

The sanitary condition of water employed for domestic purposes is, says M. Gerardin, intimately related to the presence or absence of dissolved oxygen, and the proportion of this gas present and dissolved determines the hygienic state of the water. Unfortunately, the quantitative determination of dissolved oxygen is very difficult. The French chemist just mentioned employs for this purpose hydrosulphite of soda; Prof. Wurtz employs a solution of pyrogalline acidified with hydrochloric acid. Gerardin has discovered that when water retains a normal proportion of dissolved oxygen, the lives of fish and green plants are preserved. As the oxygen diminishes, those animals which have the most active respiration first disappear, and sub- Dissolved oxygen.

sequently those of lower respirative powers; and he concludes that organic matters in a state of decomposition deprive water of its dissolved oxygen, and consequently render it impossible for either plants or animals of superior organization to live in it.

Amateur water analysis. A person with no knowledge of chemistry cannot commonly, if ever, be trusted to determine the potability of water, or judge of the comparative excellence of two or more kinds of water, provided they do not contain impurities which are *Deceptive characteristics.* plainly palpable to the senses. There is a great deal of difference in the color and taste of potable waters, and one not an expert in judging of water might be misled in many instances—especially as some of the most dangerous impurities occurring in natural waters are less readily detected than those which, though harmless, noticeably affect their color and taste. I have seen and drank very wholesome and satisfactory water—notably that drawn from the juniper swamps of Virginia—which one not accustomed to would hesitate to taste; I have also seen water drawn from deep wells, cool, clear and sparkling, which was the means of poisoning many people—some of them fatally—and which no one could drink with impunity. The notion that taste and smell can be relied upon as sure indications of the quality of water is a dangerous error. In 1866 an epidemic of cholera was caused in London by the use of impure water from wells poisoned by foul liquids permeating the soil; and yet we have it on the authority of Dr. Letterby and other eminent sanitarians and chemists, that many of these wells "yielded cool, bright and clear water." In the sixth report of the Rivers Pollution Commission it is asserted that samples of water taken from London wells consisted almost wholly of soakage from sewers and cesspools, and that some of them actually possessed a manure value one hundred and fifty per cent. greater than that of an equal volume of London sewage. Many of these polluted wells enjoyed a high reputation, and water drawn from them was considered better than that furnished by the water companies.

WATER SUPPLY IN COUNTRY DISTRICTS. 299

For these reasons it is commonly advisable, though not always necessary, for those selecting sources of permanent water supply in the country to refer the question to a chemist experienced in water analysis. A qualitative analysis will usually answer, and the small expense attending the employment of a chemist for this purpose will be found a good investment. *Advantages of employing a chemist.*

The sources of water supply in country districts are wells, springs, and, occasionally, ponds and running streams. Rain water is also a dependence for many uses on account of its softness. Of these several sources of supply, wells are given a decided preference in this country. A well is a shaft by which we reach the water-bearing stratum, the depth depending upon the formation through which it is sunk. They are of three kinds—those which are dug, those which are bored (commonly called artesian wells), and those which are made by forcing a tube into the ground (commonly called drive or driven wells). These will be considered separately and at such length as the scope of this chapter may warrant. *Sources of water supply. Wells.*

The common open well, excavated by digging, dates its origin back to the remotest antiquity. Some of the oldest wells in the world are among the most remarkable, extending to a great depth through solid rock, with winding pathways, also hewn in rock, descending to the water level. In some instances these pathways are so wide and so easy in their descent that one may ride down to the water on horseback. Joseph's well, at Cairo, Egypt, is 297 feet deep; Jacob's well, near Sychar, formerly known as Shechem, on the road to Jerusalem, is 105 feet deep. We have many wells much deeper than these, but none so remarkable when we consider the primitive character of the tools employed in digging them, the nature of the strata through which they pass, and the fact that the ancients had no explosive agents for blasting. *Open wells. Remarkable ancient wells. Joseph's well. Jacob's well.*

As the work of well digging is commonly performed by men specially skilled in the art, and as the methods and tools employed are of little interest to the general reader, I shall limit *Well digging.*

my remarks to a few suggestions respecting the location and care of wells.

Water-bearing strata. In most localities desirable for residence, and in many which are not, water can usually be obtained within reasonable distance of the surface. It is not commonly necessary to go more than 20 or 30 feet to reach a water-bearing stratum. In exceptional instances, as in cases where wells are sunk in gravelly knolls or in other positions where the local formation is such as to bring the water-bearing stratum further from the surface, it is necessary to go deeper. Under these circumstances wells *Location of wells.* are sometimes as much as 80 feet in depth. In selecting a site for a well we must remember the fact that a hole in the ground, for whatever purpose made, is very liable to become under ordinary circumstances a receptacle for surface drainage, as well as a cesspool for other matters which find their way from *Sources of contamination.* the surface down through permeable strata. For this reason wells should be as far as possible removed from barn-yards, muck-heaps, privies, cesspools and other possible sources of contamination. *Earth not a good filter.* The idea that simple filtration through soil will free water from organic impurities and rob it of all hurtful properties, is clearly erroneous. Adequate filtration will do this for a time, but the fissures and veins through which surface water makes its way down to the strata from which wells are supplied, may become so filled with impurities that they would poison pure water passing through them. It is not unusual for wells to become suddenly affected from causes which have been operating without apparent effect for years, and the germs of disease are thus often taken into the system under circum- *Organic impurities.* stances which excite no suspicion. Decomposing organic matter of any kind, and especially animal excrement, is the great source of danger, and when such matter is allowed to accumulate in the neighborhood of wells upon the surface of any but the most exceptionally impervious soils, we may be very certain that sooner or later it will render the water impure and un- *Local sources of contamination.* wholesome, if not positively fatal. It is not uncommon, however, to find wells in light sandy or gravelly soil in close proximity to privy vaults, cesspools and barn-yards. They are, as the rule, lined with loose stones which afford surface water free access, and are often, especially in times of heavy rain or when the winter snows are thawing, receptacles for water charged with all manner of impurities scoured from the surface. I have seen instances in which a well, barn-yard, garbage heap, a pervious cesspool receiving the liquid refuse of the house, a privy, a house and a well-manured garden, were all to be found within the limits of a plot 50 by 150 feet, and in soil so light that only after heavy and long-continued rain would any water be found standing on the surface. Under such circumstances it is scarcely possible for the water in a well to be fit for use. In calling attention to a dangerous proximity of well and privy, or cesspool, I have usually found people unwilling to believe that any harm could result therefrom—or indeed from any other source of surface contamination provided the earth was banked up 6 or 8 inches around the curb.

In villages and towns where the dwellings are close together and wells are necessarily in close proximity to houses and outbuildings, the danger of water poisoning is imminent at all times. That such villages and towns are commonly healthy places of residence in this country would prove nothing if true. In Great Britain some of the most fatal epidemics of typhoid fever have occurred in such towns, and the same is true in this country. As the rule our towns and villages have outgrown the simple methods and appliances that provided adequate drainage for the few comparatively isolated houses forming the nuclei around which they have gathered. Localities which were conspicuously healthy half a century ago, and which for that very reason attracted population, are now scourged with malaria, and not a few of those who can do so are removing to the cities to regain lost health. There is probably not a village or settlement within 40 miles of New York in any direction that is free from malarious diseases of local origin. There may be particular neighborhoods to which these remarks do not apply, and which, from natural advantages of location, with good drainage and abundant water supply, have the means of keeping healthy; but most of the country which has afforded accommodation to the population crowded out of New York by high rents and the insufficiency of home accommodations within its limits is not healthy, and cannot under existing conditions expect to be. In a majority of cases defective and insufficient drainage, with its attendant and almost inseparable evil of water poisoning, is at the root of the whole trouble. In but few instances have attempts been made to deal with these evils by measures of sanitary reform. Each house has within the little inclosure it occupies its own source of water supply and its own sewer—the former a well and the latter a leaching cesspool. Even when the true nature of the evils undermining the public health is understood and appreciated, it is not always possible to remedy them. It is easy to levy a tax for such improvements as stone sidewalks and road macadamizing, but a tax for draining a village and supplying it with water finds few advocates, even among those whose families have suffered most from diseases which such reforms would forever banish from the neighborhood. It is a well-established fact that since the introduction of aqueduct water into large cities, typhoid and many other fevers, resulting among other causes from water poisoning, have become more common in the country than in cities. This I have no doubt is attributable to the poisons taken into the system in well water contaminated with organic matter. Instances of such contamination which have given rise to zymotic diseases are numerous and well authenticated. The slimy matter often found covering stones in wells is a true fungoid growth, and is an active blood poison when taken into the system. In wells polluted by sewage these fungoid or confervoid growths are, I believe, always found.

The danger of impure water is a subject to which public attention cannot be too often nor too forcibly called, and I regret that the scope and purpose of this volume prevents the colla-

WATER SUPPLY IN COUNTRY DISTRICTS. 303

tion and presentation of the mass of exact and unquestionable evidence which establishes the fact that water poisoning is a fruitful cause of disease and death. Where wells are necessarily depended upon, the only means of protecting them from contamination which admit of general application are, I believe, those suggested in the preceding chapter.

There should be no permanent woodwork inside a well. It attracts and shelters various forms of animal life, and by its decay, which is likely to be rapid, it may give rise to unwholesome conditions affecting the character of the water. It is also desirable that the curbing should be of stone or brick, and it is always a good plan to cover open wells with low, wide-spreading roofs of steep pitch, supported on corner posts and inclosed on three sides with lattice work. Such roofs, if kept clean and in good repair, will be useful in excluding dust, leaves and other matters likely to affect the quality of the water. Anything falling into a well should be fished out at once, and all wells should be cleaned as often as may be necessary. Some wells foul more rapidly than others, especially those shaded by overhanging trees which drop rain and shed leaves into them. The condition of the sides and bottom of an open well can always be determined on a bright day by reflecting the sunlight into it with a mirror. *Woodwork in wells. Curbing. Covers. Cleaning wells.*

When the character of the formation is such that good water cannot be reached by open wells of convenient depth, recourse is commonly had to the boring of what are known as artesian wells. These, though costly, are often of great advantage in securing an abundant supply of water in localities where it cannot be had at or near the surface. Many wells of this kind have been bored in this country within the past few years, some of them to a great depth, and the proportion that have failed to give an abundant and unfailing supply of water is not great. The objections to artesian wells are their great cost and the uncertainty of striking water of the right kind. At the time of this writing there is an artesian well at Reading, Pa., *Artesian wells. Uncertain results of deep boring.*

2000 feet deep, costing $22,000, which contains 47 grains Epsom salts to the gallon. An artesian well at Fifth and Cherry streets, Philadelphia, containing 116 grains of foreign matter to the gallon, can only be used to condense steam for the boiler. An artesian well in South street, Philadelphia, furnishes water not fit for steam. In the suburbs of the same city there are two artesian wells, each 100 feet deep, but the water of both is so impure that it can only be used for condensing. The water of the well at the Continental Hotel is not pure. At Louisville there is a well 1649 feet deep, one in St. Louis 2080 feet deep, one in South Bend and one in Terre Haute, but the water of them all is impregnated with minerals and fit only for medicinal uses. At Atlantic City a number of wells have been bored in the hope of getting pure water, but not one yields water fit for household use.* Instances have been known where the sinking of a second artesian well near one which was yielding an abundant supply of water, has not only proved a failure, but has for some cause spoiled the first success. Such accidents,

Favorable conditions. however, are exceptional. The rocks of the paleozoic series extend in nearly horizontal strata over the greater part of North America, and the geological structure of the continent is thus particularly favorable to the general employment of arte-

Expense of boring. sian wells as sources of water supply. As the boring of these wells is a special industry, practiced by few and commonly under the direction of engineers, they are beyond the means of most individuals residing in suburban or rural districts. This would not be the case, however, were such wells common

Wells in Artois. enough to encourage an active competition for orders. In the province of the Artois, France, from which these wells are named, the use of boring tools has been practiced for several centuries, and although the apparatus employed is of the simplest and rudest kind, such wells are to be found beside the door of nearly every cottage.

* The statements made in the few preceding lines respecting artesian wells are based upon information gathered during the winter of 1875.

The artesian well seems to afford, in many instances, the only *Special utility of artesian wells.* solution of the problem of supplying with water small towns which, though not large enough to justify the expense of public water works, are too crowded to render open wells extending only to the first water-bearing stratum a safe dependence. Such wells, as has been already shown, are continually liable to become contaminated by surface water, by house drainage and by foul liquid absorbed by the soil. It is usually difficult, and *Town and village supplies.* often impossible, for town and village authorities to find within practicable distances sufficient supplies of water of suitable quality for home service so situated as to be available. In too many instances the water question has been decided without due consideration, and, as a consequence, many of our towns are supplied with water which the sanitarian cannot but regard with suspicion and disfavor.

A cheap modification of the artesian well is the drive well, *Drive wells* made by driving into the ground a metal tube perforated at the end and armed with a sharp point. These wells have become very popular in this country and are used to a considerable extent abroad, and they possess certain important advantages over the old form of open well. The principal objection *Galvanized tubes.* to wells of this class is found in the general use of galvanized iron tubes. No important advantage is gained by coating a pipe with zinc which cannot be better secured by other and safer methods of protecting the iron against rapid oxidation. The great merit of the drive, or driven, well consists in the *Advantages of drive wells* ease and rapidity with which it may be sunk. In light, open soils, half an hour usually suffices to strike water and have a pump going. These wells average in depth with the usual form of open wells. They are easily protected against direct contamination by the inflow of unfiltered surface water, but are as liable as open wells to be rendered unwholesome by organic impurities which penetrate the soil and find access to their sources.

In many parts of the country it is possible to obtain from *Springs.* springs a sufficient water supply without the trouble and
20

expense of digging deep wells. When unfailing springs of good water are at hand, nothing more is to be desired. It does not follow, however, that because the water in a spring is clear and cold and free from any unpleasant taste, it is fit for drinking and household use. As already stated, many dangerous impurities are not apparent to the senses, unless, possibly, under the microscope. As springs are altogether fed by water which has fallen upon the surface, and as the purity of the water they contain depends upon the filtrative power of the superficial strata through which it passes, it is extremely liable to become contaminated if its course lies through any impurities. In its passage through the lower strata of the atmosphere, water discharged from the clouds as rain absorbs oxygen, nitrogen and carbonic acid (the amount averaging about 6·93 cubic inches to the gallon), divided about as follows: Nitrogen, 64 per cent.; oxygen, 34 per cent.; carbonic acid, 2 per cent. It is also customary, especially in populous districts and near cities and large towns, to find a trace of ammonia, which is commonly combined with nitric, sulphuric or carbonic acid. These dissolved gases are usually held by water in its passage through the earth to the reservoirs whence springs are fed, and others are taken up, as well as various soluble mineral impurities, from the earth itself. The percentage of dissolved gases is from two to five times greater in spring water than in rain water, those of most common occurrence in the former being oxygen, nitrogen, carbureted hydrogen, sulphuretted hydrogen and carbonic acid. The mineral impurities taken up by spring waters are so numerous, and differ so widely under different circumstances, that it is difficult to make the list complete. The minerals of most frequent occurrence are lime, magnesia, potash, soda, iron and manganese. These are usually found as sulphides or chlorides. To what extent the quality of the water, as regards its fitness for domestic use, is affected by these impurities, depends upon circumstances. What has already been said in preceding pages regarding water analysis is probably sufficient for the non-professional reader. Generally speaking, good springs will yield soft, clear water, free from visible impurities and pleasant to the taste. The character of the water can be determined by the methods already described. *Characteristics of good spring waters*

When spring water is used for house service or for drinking, it is important that the spring from which it is drawn should be frequently and carefully looked after. It should be kept clean and free from rank vegetation, and protected against contamination by water which has not been filtered through the soil. Cattle and horses should not be allowed to trample the soil about it, or they will be very sure to pollute it. In other words, the spring should be as well taken care of in its way as the water pitcher which comes to the table. I know an instance in which a family of refined tastes were for years content to draw all the water they used for drinking and cooking from a spring choked with rank weeds and mosses, full of reptile life, and lying in the middle of a foul mud-bed, kept soft by the constant tramping of horses and cattle which were allowed free access to it. They had apparently never given the condition of the spring a minute's thought, and even when its bad condition was discovered by a sanitary inspection of the premises which followed the outbreak of diphtheria among the children of the family, it was difficult to make them understand the necessity for reform. *The care of springs.* *A neglected spring* *Indifference to known danger.*

What I have said of springs applies in a general way to ponds, running streams and all other available sources of water supply for country houses. In the case of streams and of ponds fed by streams, it is important to trace them to their sources. If a stream flows through swamps or low-lying pasture lands, or receives the drainage of swamps, stagnant pools or dwellings, it is fair to assume that safer and purer water can be had by digging or boring. If the supply is drawn from a pond, it must be kept clean and free from accumulations of fallen leaves, decaying wood, organic impurities, &c. When *Ponds, streams, &c.* *How streams are fed.* *Cleaning ponds.*

the importance of pure water is better understood and appreciated by the public, and proper means are taken to secure it by all who are able to do so, there is reason to believe that the death rate in country districts will show a marked falling off from present averages.

Expedients for the purification of water. It is sometimes impossible for the temporary resident in a country house to obtain satisfactory drinking water. Even though he knows what should be done to secure a supply of good water, he cannot always apply his knowledge in carrying out necessary reforms, and must perforce take things as he finds them. In such cases he can usually neutralize the more dangerous impurities if he will take the trouble to do so. The dangers attending the use of impure water are so great, however, that it is scarcely safe to give the information. I would impress it upon the mind of the reader that methods of rendering impure water fit for drinking should only be resorted to in case of extreme emergency. In all other cases total abstinence is the best and indeed the only safeguard, as none of the methods which can be employed for rendering it less dangerous can be absolutely depended upon.

Danger of impure water.

Uncertainty of purification.

Methods. The means by which water is purified and rendered potable may be briefly summarized as follows:

Filtration through charcoal or iron sponge.

Oxidation of organic matter by permanganate of potash.

Precipitation of organic matter with sulphate of iron or alum.

Destruction of organic germs by boiling.

Charcoal filters. In all cases of suspected impurity the filter should come into use, for fresh charcoal possesses a remarkable power of arresting and retaining not only suspended, but even dissolved metals and salts. A good charcoal filter, for use on a large scale, is made of bone-black ground to a fine powder and mixed with Norway tar and other combustible materials, also in a fine powder. The mass is kneaded into a plastic condition with liquid pitch, moulded into blocks and exposed to a great heat. This produces a very porous block, the pores do not so soon become clogged, and all the water comes in contact with the charcoal. The value of iron as a purifying agent has been mentioned in a previous chapter. Iron sponge, when used as a filter, renders important service in oxidizing organic impurities, but it is dissolved slowly and carried away in the form of a soluble carbonate; hence it must be followed by another filter to hold this back—say one of marble dust. *Iron sponge filters.*

For the partial purification of water on a small scale the most convenient agent is, probably, permanganate of potash. This should be added until a faint pink tint is seen in the water. Time is then allowed for visible impurities to settle and the water is decanted for use. This has been recommended as a certain method of freeing water from organic impurities, but it is by no means a sure dependence. According to Henri de Parville, animalculæ remain in full life and vigor in water to which permanganate of potash has been added in large proportion. Hence, too much dependence should not be placed upon it. *Permanganate of potash.* *Vitality of animalculæ in water.*

For the purification of water on a large scale the neutral sulphate of peroxide of iron may be used with good results. The proportion in which this solution is to be added to the water is determined by the degree of impurity to be removed; and the proportions suitable must therefore be determined by careful experiment, practiced from time to time if the impurity of the water is found to vary. The water to be purified may be run into a tank or reservoir, and the solution of neutral sulphate added as it runs in, so that it may be well mixed with the water. A short time after the neutral sulphate is added to the water it becomes decomposed, and forms, with some of the impurities contained in the water, a basic salt which is insoluble in water. The solid and insoluble particles of this new salt are precipitated, and, together with the impurities contained in the water, form a sedimentary deposit, from which the purified water may be allowed to run off, leaving *Neutral sulphate of peroxide of iron.*

the deposit in the tank or reservoir. A repetition of this precipitating process on other bodies of water which may be run into the same tank or reservoir, will cause additional deposits, which, when allowed to accumulate to a sufficient depth, may be collected and removed from the reservoir from time to time.

Boiling. Whether the boiling of water is efficacious as a means of destroying animal and vegetable organisms and their germs, has been much discussed among those claiming to be authorities on the subject. It is enough to say that there is no difference of opinion as to its being the best and surest way at this time known of effecting the purification of water containing *Destruction of animalculæ* organic matter. Animalculæ are not salamanders, and after we have destroyed their life we can, if we like, complete the process by cremating their corpses with permanganate of potash, precipitate them with alum or iron, or strain them out with a charcoal filter.

Restoring the flavor of boiled water. After water has been boiled it should be cooled for 24 hours or longer in a cold cellar, to restore its "freshness" and remove the "flat" taste. It is impossible, however, to make it as agreeable to the palate as water which has never been *Why boiled water is "flat."* boiled. The reason for this is that boiling has expelled all the free carbonic acid, besides decomposing the bicarbonates of the alkalies and alkaline earths, leaving the monocarbonates *Unpleasant taste and smell.* behind. Another reason for the disagreeable taste which often belongs to boiled water is found in the fact that some of the organic matter is decomposed, partly by heat and partly under the influence of the carbonated alkalies, giving rise to compounds of unpleasant taste and smell. When the character of the organic impurities is such that they make the water disagreeable, it is probable that the difficulty will be remedied by adding an acid before boiling. To restore the free carbonic acid and render the water "lively," bicarbonate of soda may be employed.

Citric acid. The best acids to employ are probably citric and hydrochloric. If the former is used, care must be taken not to subsequently add enough bicarbonate of soda to make the water neutral to test paper, as that would prevent any appreciable improvement in the flavor. The taste is a safe guide. Enough acid is added to give the water a sharp and unpleasantly sour taste; the water is then boiled, and after cooling neutralized until the unpleasant acidity has given place to the refreshing taste of the acid salts. When hydrochloric acid is used, less than 58 grains to the gallon is amply sufficient, but it is of great importance that it should be free from thallium and arsenic, nitric acid, chlorine and sulphurous acid—impurities often found in the hydrochloric acid of commerce. For this and other reasons the process cannot be recommended unless employed under the direction of a chemist. Citric acid is much simpler and probably safer. *Method of employment. Hydrochloric acid. Importance of purity.*

The taste of water not positively offensive after boiling can usually be sufficiently disguised to render it agreeable to most palates by pouring it upon tea leaves. Cold tea, when not too strong, is an agreeable and refreshing beverage, and may be used moderately without hurtful results. I should doubt its advantages as a steady drink, but should at any time prefer it to unboiled, impure water. Iced tea has become a very popular summer beverage, and while usually taken much stronger than is necessary to disguise the flavor of water, it is often preferable to ice water. *Tea as a flavoring for boiled water.*

As I said in the introduction to the concluding remarks of this chapter, the best precaution against injury from impure water is not to use it. No one can afford to be indifferent to the dangers of water poisoning, and when hurtful impurities are known or suspected in water, no trouble or expense should be spared to secure a better supply. When this is impracticable for any reason, the person compelled to use impure water may have temporary recourse to whichever of the expedients for purifying it suggested in the preceding pages he may find most convenient. *Abstinence from impure water the best safeguard.*

CHAPTER XII.

SUGGESTIONS CONCERNING THE SANITARY CARE OF PREMISES.

It has not been the author's intention in writing this book to make it a manual of sanitary science, but a few remarks on the sanitary care of premises will not be out of place, as it not infrequently falls to the lot of those for whom this work is intended to deal with unhealthful conditions existing outside of the pipe systems of houses.

Cleanliness. The first essential condition of healthfulness is cleanliness. The shovel, the broom, soap and water, sunshine and ventilation, are the agents upon which we must mainly rely in guarding against unhealthful conditions in our surroundings. How, when and where the broom, shovel and scrubbing brush need *Dirt.* to be employed, the reader must decide for himself. I can only say in a general way that anything and everything which can be properly classed as "dirt" should be put where it belongs. It will then cease to be dirt. There are few things so dangerous that we cannot rob them of their power for mischief *Decaying organic matter.* by putting them in their proper places. Decaying animal and vegetable matter of all kinds should be carefully composted and used for manure. We thus return it to the earth where it belongs, and its elements remain in the soil and are taken up and assimilated by vegetation. Under these circumstances decaying organic matter fulfills its ultimate functions, and in so doing is powerless for harm. This of course applies especially to country houses. In cities there is neither opportunity for composting nor use for manure; but the conditions are also different, in that the occupants of city and town houses can *Neglect of health in cities.* usually dispose of all organic refuse without difficulty. Unfortunately, however, we often see in cities a gross neglect of sanitary laws, resulting in great part from the ease with which

SANITARY CARE OF PREMISES. 313

filth can be got rid of. The foul, offensive ash barrel standing unattended to for hours under our parlor windows, and the sour and sickening swill pail perfuming our back areas or awaiting on the curbstone the arrival of the scavenger's cart, are evils peculiar to cities, and, I think, more noticeable in New York than anywhere else. I have often suffered severe nausea when walking through elegant streets in the upper wards of New York, caused by the horrible smell of ash and garbage receptacles strung along the sidewalk about ten paces apart. As the rule these vessels are barrels or firkins, which readily absorb and retain a share of the foulness they contain, and which are rarely cleaned out or disinfected. In most families a barrel once dedicated to this ignoble service remains in use until it falls to pieces from rottenness or is fortunately stolen by street gamins to feed election-night bonfires. During ten hours of the day these receptacles stand beside the kitchen doors or under the front stoops. They are set out at nightfall on the curbstone, emptied when the ash and garbage men come around, are by them replaced on the curbstone and, at the convenience of the housekeeper or servant, reclaimed. Many houses, especially those occupied by several families, have large permanent ash bins on the sidewalk. These are wooden boxes with one side cut down to make it more convenient to partially empty them with a shovel. They are seldom or never cleaned, and as they are receptacles for more different kinds of nastiness than could be named, they soon become in summer disease-breeding nuisances. They are only worse than the typical ash barrel because larger.

<small>Ash, swill and garbage receptacles.</small>

<small>Ash barrels.</small>

<small>Garbage bins.</small>

The only vessels suitable for this service are those made of galvanized iron. As they are emptied every twenty-four hours, there is no excuse for the offensiveness which almost invariably characterizes them. There are but few things in the waste of a house which enter the garbage receptacle in a state of decomposition, and when such a receptacle stinks it is evident that it is neglected. An occasional—if necessary a frequent—scalding

<small>Galvanized-iron garbage receptacles.</small>

out, followed by a thorough scrubbing with a broom and an airing in the sun, will correct any tendency to offensiveness in a metallic vessel, and go far toward reforming an evil the existence of which is a perpetual surprise.

Sanitary care of cellars. The sanitary care of a house should in all cases extend to the cellar. In a previous chapter I had something to say on the subject of country cellars. The same remarks apply with even greater force to the cellars of city dwellings. As the rule these *Refuse in cellars.* are neglected. Vegetables not fit for use are allowed to remain in them and decay; dirt of all kinds accumulates in dark corners; coal dust, always damp, gives off sulphurous gases which are peculiarly irritating to the throat and lung passages; mold gathers on floor and walls, and foul and unhealthy conditions exist in every part. When such cellars are cleaned out the amount of rubbish and dirt removed is a ten days' wonder to the householder, but he consoles himself with the reflection that it will not again need cleaning for a very long time. Considering the condition in which a large proportion of city and town house cellars are allowed to remain from one year's end to another, there is no occasion for the surprise often felt at the mortality tables and the prevalence of diseases which have no excuse for being.

Disinfection. When everything has been done in the way of cleansing and purification which is possible with broom, shovel, clean water, fresh air and sunshine, it is sometimes necessary to have recourse to disinfection as a means of correcting unwholesome conditions. As there exists a very general misapprehension outside of the medical profession with regard to what are known as disinfectants and their uses, some remarks on this subject may have interest and value for practical people.

Prof. Baker on disinfectants. In a letter from Prof. II. M. Baker, a chemist who has given much attention to this subject, addressed to Surgeon II. M. Wells, of the U. S. Navy, the theory of disinfectants is so well presented that I cannot do better than quote therefrom. Prof. Baker says:

SANITARY CARE OF PREMISES.

"As the action induced in the process of 'deodorizing,' 'disinfecting,' &c., varies according to the agent employed, it is impossible to make a general rule applying to all substances possessing such characters, but one may acquire a general knowledge of their mode of operation upon special well-known principles.

"It is a theory of chemistry that any body of organic constitution (especially if one of its elements be nitrogen) is subject to enter into spontaneous decomposition under mild influences, such as a certain range of temperature, the presence of moisture, the action of direct or diffused light, or contact with another body of like feeble structure. The reason for such properties is founded upon the fact that, for the most part, the greater the number of chemical elements existing in combination to form a particular body the more feeble becomes the chemical affinity that compels such combination, and should nitrogen be one of those elements then the chemical constitution is rendered very much less stable, because nitrogen is very feeble in all its affinities. Those bodies which emit foul odors are of organic structure, and it is during the progress of what we call 'spontaneous decomposition' that these odoriferous compounds are evolved; so that any substance placed in contact with the decomposing matter, which arrests chemical dissolution or putrefaction by displacement, substitution, elimination, direct combination, mutual decomposition, or by inducing a change of molecular structure, or catalysis, thereby forming a new and stable compound, may justly be styled a 'deodorizer.' *Decomposition of organic matter.*

"The bodies of most frequent occurrence, and that exist in excessive quantities, which exhale offensive odors during decomposition, are the animal and vegetable tissues—as albumen, gelatine, fibrine, caseine and a vast number of nitrogenous compounds from the blood, bile and excrementitious substances at slaughter-houses and provision stores, and also fæcal and urinary matter in water-closets, urinals, &c., besides the unexamined products of decomposition in cesspools, sinks, offal barrels, casks and the like. *Offensive products of decomposition.*

316 SANITARY CARE OF PREMISES.

Infections and contagions. "The term 'disinfectant' is often employed as though it indicated or implied anti-infection, but it seems that its meaning might with propriety be extended to substances which induce the chemical destruction of, or removal from, infected tissues of virulent matter. Some infections are of local and others of a general character and may be communicated by contact, but many contagions are supposed to be transmitted by the atmosphere to the lungs, where the poisonous matter meets the blood, and thereby finds food which it appropriates to its own growth, against the faithful protest of the vital powers.

Power of a virus. "The power of virus is chemical in its character; so if the vital forces are in a depressed condition it is most probable the chemical forces will acquire the mastery, although an active strife exists between the two. No positive knowledge prevails as to the origin of infecting bodies, nor any proofs of distinct characteristics except in the effects manifested. The venom of the reptile differs from the virus of rabies and variola, and these three again from the carcinoma of the cancer.

Anti-infectants. "An 'anti-infectant' cannot be indicated through the aid of reason until a sufficient quantity of the isolated contagious matter can be procured for the investigation of its properties; so we must content ourselves with the employment of those agents which experiment and observation proclaim most trustworthy.

Disinfectants. "A 'disinfectant' should possess the property of destroying the chemical structure of virus and thereby produce in its stead a body with inert characters, and consequently afford the natural chemical and vital forces an opportunity to pursue their regular vocation or function of removing effete matter, and replenishing the exhausted tissues unobstructed."

Efficacy of disinfectant preparations. Concerning the efficiency of the various disinfectants available for general use it is difficult to speak with confidence in all cases. The opinions of experts in sanitary matters seem to be undergoing a gradual change, as conclusions formed in the laboratory are contradicted, modified or confirmed by practical experience, or *vice versa*. I can only give my own opinions, formed in part from practical experience in the use of many of the disinfectants generally employed, but still more from a study of the results obtained by eminent and trustworthy experimenters.

What are known under the general name of disinfectants may be classified as follows:

1st. Positive disinfectants, which destroy or restrain infectious virus. <small>Positive disinfectants.</small>

2d. Antiseptics, which merely prevent or arrest fermentation and decay. <small>Antiseptics.</small>

3d. Deodorants, absorbents, &c., which destroy bad smells, deodorize putrid exhalations, or absorb moisture and gases. <small>Deodorants.</small>

As the functions of these several chemical preparations are quite distinct, it is important that, before selecting one for use under given conditions, we should know just what we want to accomplish. In certain cases two or more kinds of disinfecting material may be advantageously combined, but such admixture must be made by persons familiar with their properties or uses or there is danger of defeating the object sought by bringing in contact substances which neutralize each other and together become inoperative.

Carbolic Acid.—This is one of the most generally used of the positive disinfectants, but it has lately fallen into some disrepute among chemists. It is probable, however, that the laboratory tests have not in all cases been made under conditions which determine its general value for use in the sanitary policing of premises. Employed in solution, carbolic acid is a powerful destroyer of the lower animal organisms. Solutions of one part to 2000 parts of water kill infusoria and bacteria instantly; 2 milligrams (0·03 grains) will stop fermentation in 100 c. c. of sugar solution. In another experiment where meat was kept under water, in three days it was very turbid and there were numerous bacteria. In a solution of one part carbolic acid to 10,000 parts of water, the meat began to decompose in six days; but in a solution of one part in 2000 it did <small>Carbolic acid.</small> <small>Its powers and methods of employment.</small>

not putrefy until the expiration of five weeks, when all the acid had evaporated. In water containing 1 per cent. of carbolic acid, at the end of eight weeks the meat had the appearance of fresh meat, and no bacteria could be detected. From a practical point of view there is probably nothing so well adapted as carbolic acid to prevent the evolution of putrefying and infectious organisms in large masses of readily decomposing matter (such as excreta and sewage), until they can be removed and rendered harmless in other ways.

For employment as a disinfectant it may be diluted by adding from 40 to 100 parts of water to one part of acid. It may then be sprinkled upon garbage or decaying organic matter, upon unclean surfaces, in drains and elsewhere with good results. A more perfect solution of carbolic acid in water is secured by adding one part of strong vinegar to one part acid. This is only necessary or desirable, however, when clothing or other fabrics are to be washed in the solution.

Coal tar products. Both carbolic and cresylic acids are coal tar products, and are not properly acids but alcohols. Coal tar itself is useful in many ways as a disinfectant, and if mixed with sawdust or dry lime may be employed with advantage in foul places or upon heaps of decaying refuse.

Sulphate of iron. *Sulphate of Iron and Sulphate of Zinc.*—These salts possess well-known properties as positive disinfectants. The former is expressly recommended, in admixture with carbolic acid, for the disinfection of privies, cesspools, drains, sewers, and all vessels or places where discharges from the sick are evacuated. The sulphate of iron (copperas) is dissolved in water in the proportions of eight or ten pounds of the former to five gallons of the latter, and half a pint of fluid carbolic acid is added. In using this preparation for the disinfection of privies, sewers, drains or garbage heaps, pour in or throw on the solution, a pint or so at a time, about once an hour until the nuisance is corrected.

Sulphate of zinc. Sulphate of zinc solution is made by dissolving two ounces of the salt in a gallon of water. It is chiefly useful for disinfecting clothing, bedding, &c., and answers as a temporary expedient until they can be more thoroughly cleansed by boiling.

Permanganate of Potash.—As a disinfectant, permanganate of potash is of very little value, although favorably regarded by some. Schroter has observed that infusoria swim around for a long time in the strongest solutions of this substance; then these organisms turn brown inside and die. Yeast cells act in a similar manner, while the spores of mold fungi germinate even in the strongest solutions. Bacteria are killed in concentrated solutions without turning brown; in solutions of 1 to 1,000, on the contrary, they increase. The effect of permanganate is still further weakened by first acting upon decomposed organic matter, and is thereby decomposed. For example, if a piece of fresh meat is put into a solution of permanganate of potash, its surface becomes brown, the solution is soon decolorized and the permanganate is decomposed. The water then extracts substances from the undecomposed meat; bacteria appear, multiply rapidly, and the meat is further attacked. On account of the large mass of decomposed organic matter, a great quantity of permanganate is required for its repeated disinfection, and still after one or two days there is a large increase of bacteria—turbidity and the odor of decay reappearing. In spite of the use of large quantities of this disinfectant, the meat spoils almost as rapidly as in pure water. Permanganate of potash may, therefore, be used with profit for washing out wounds, but for disinfecting filth it is totally unsuitable.

Quicklime and Chloride of Lime.—As an absorbent of moisture and putrid fluids, quicklime is good. Fresh stone lime should be broken into small pieces—the smaller the better—and sprinkled on the places to be dried. Sick rooms, stables, outhouses, &c., may be greatly purified by whitewashing them with lime, but little or no benefit is derived from kalsomining. The free use of lime may be recommended, but the admixture of whiting, chalk, glue or any other of the substances usually employed to improve the whiteness or insure the adhesion of lime, will do much to neutralize the benefits of limewash.

Chloride of lime. The value of chloride of lime is doubtful. It gives off chlorine, which is supposed to form stable compounds with the products of decomposition in bodies containing nitrogen, but it is **Chlorine gas.** doubtful if the resulting compounds are at all stable. Furthermore, dry chlorine gas has no effect upon the lower organisms, and it is totally useless to fumigate clothing with chlorine. Its disinfecting action upon liquids, excretia, and the like, is insufficient, and very rapidly exhausted.

Charcoal. *Charcoal.*—As an absorbent of foul gases and a general purifying agent, charcoal is invaluable. It should be fresh and dry to give the best results. Voelcker says of charcoal: "It possesses the power not only of absorbing certain foul smelling gases—sulphuretted hydrogen and ammonia—but also of destroying the gases thus absorbed; for otherwise its purifying action would soon be greatly impaired. It is very porous, and its pores are filled with condensed oxygen to the extent of eight **Oxygen.** times its bulk. We have, therefore, in charcoal, oxygen gas in a condensed form and more active condition than in the air we breathe; hence it is that organic matter in contact with **Pores of charcoal.** charcoal is so rapidly destroyed." Liebig says that "a cubic inch of beechwood charcoal contains pores equal in area to 100 feet." The oxygen contained in these pores attacks and burns up whatever is absorbed into them, and as the powers of charcoal are self-renewing, it will retain its properties for a very long time. It should be used whenever there are noxious exhalations to be absorbed and destroyed.

This by no means concludes the list of disinfectants in use, but from those recommended the reader can select one or more which will be of service as an aid in sanitary work. This book is intended chiefly for practical people who are not likely to venture any difficult experiments, and who would, as the rule, have difficulty in procuring other disinfecting agents than those above noted. During the past few years a number of disinfecting fluids have been introduced into use, and are usually sold by dispensing druggists. They are somewhat costly, considering the value of the materials employed in making them; but, so far as my experience goes, they are convenient for use on a small scale, and will accomplish all that any disinfectant is capable of if used as directed.

In the reports of the Medical Officer of the Privy Council and Local Government Board (New Series No. VI) for 1875, is given an elaborate report by Dr. Baxter on the experimental study of disinfectants, which is probably the most valuable of recent contributions to the literature of this subject. I have only space for his conclusions, which are as follows: *Dr. Baxter on disinfectants.*

I. Evidence has been adduced to show that carbolic acid, sulphur dioxide, potassic permanganate and chlorine, are all of them endowed with true disinfectant properties, though in very various degrees. *True disinfectants.*

II. It is essential to bear in mind that antiseptic is not synonomous with disinfectant power; though as regards the four agents enumerated above, the one is, in a certain limited sense, commensurate with the other. *Antiseptics.*

III. The effectual disinfectant operation of chlorine and potassic permanganate appears to depend far more on the nature of the medium through which the particles of infective matter are distributed than on the specific character of the particles themselves. *Chlorine and permanganate of potash.*

IV. When either of these agents is used to disinfect a virulent liquid containing much organic matter, or any compounds capable of uniting with chlorine, or of decomposing the permanganate, there is no security for the effectual fulfillment of disinfection short of the presence of pure chlorine or undecomposed permanganate in the liquid after all chemical action has had time to subside. *Their action.*

V. A virulent liquid cannot be regarded as certainly and completely disinfected by sulphur dioxide unless it has been rendered permanently and strongly acid. The greater solubility of this agent renders it preferable, *cæteris paribus*, to chlorine and carbolic acid for the disinfection of liquid media. *Sulphurous acid.*

322   SANITARY CARE OF PREMISES.

Carbolic acid. VI. No virulent liquid can be considered disinfected by carbolic acid unless it contain at least two per cent. by weight of the pure acid.

VII. When disinfectants are mixed with a liquid it is important to be sure that they are thoroughly incorporated with it; that no solid matters capable of shielding contagion from immediate contact with its destroyer, be overlooked.

Disinfecting the air. VIII. Aërial disinfection, as commonly practiced in the sick room, is either useless or positively objectionable, owing to the false sense of security it is calculated to produce. To make the air of a room smell strongly of carbolic acid by scattering carbolic powder about the floor, or of chlorine by placing a tray of chloride of lime in a corner, is, so far as the destruction of specific contagion is concerned, an utterly futile proceeding.

Virus in air. IX. When aërial disinfection is resorted to, the probability that the virulent particles are shielded by an envelope of dried albuminous matter, should always be held before the mind. Chlorine and sulphur dioxide are both of them suitable agents for the purpose; the latter seems decidedly the more effectual of the two. The use of carbolic vapor should be abandoned, owing to the relative feebleness and uncertainty of its action. Whether chlorine or sulphur dioxide be chosen, it is desirable that the space to be disinfected should be kept saturated with the gas for a certain time, not less than an hour; and this in the absence of such gaseous compounds as might combine with or decompose the disinfectant, and so far impair its energy.

Disinfection of foul matter in masses. X. When the thorough disinfection of a mass of solid or liquid matter through which a contagium is disseminated is impracticable, we should guard against giving a false security by the inadequate employment of artificial means. It is probable that all contagia disappear sooner or later under the influence of air and moisture, and that the absence of these influences may act as a preservative. When, therefore, we cannot advantageously or effectually supersede the natural process of decay, we must be sure that we do not hamper it by the injudicious use of antiseptics.

XI. Dry heat, when it can be applied, is probably the most efficient of all disinfectants. But, in the first place, we must be sure that the desired temperature is actually reached by every particle of matter included in the heated space; secondly, length of exposure and degree of heat should be regarded as mutually compensatory factors within certain limits. *Heat*

In conclusion Dr. Baxter says: "The above statements are not so discouraging as they may appear at the first glance to our reliance upon artificial disinfection. If we believe that all contagia are generated, like those of small pox and scarlet fever, in the infected organism, and there only, the outlook is a hopeful one. We might even anticipate an approach to the perfect fulfillment of the work of disinfection by submitting all matters, immediately after their removal from the affected person and before any dilution or admixture, to the full influence of one or other among the destructive agencies at our command. On the other hand, if the contagium of any disease is capable of being generated *de novo*, outside the body (pythogenic origin of enteric fever, typhus created by overcrowding) such contagium can hardly be eradicated by any method of artificial disinfection. For cases of the latter kind the opening words of the memorandum previously referred to furnish the only solution: 'It is to cleanliness, ventilation and drainage and the use of perfectly pure drinking water, that populations ought to look mainly for safety against nuisance and infection. Artificial disinfectants cannot properly supply the place of these essentials, for, except in a small and peculiar class of cases, they are of temporary and imperfect usefulness.'" *Dr. Baxter's conclusions.* *Generation of contagium.* *Conditions of health.*

A committee appointed from the St. Petersburg Medical Academy by the Russian government to investigate the same subject (antiseptics and disinfectants), arrived at the following conclusions: *Russian report on disinfection.*

1. Carbolic acid is the most efficient means against the development of ammoniacal gas, putrescence and development of lower organisms in organic matter under decomposition, and is, therefore, the best antiseptic. 2. Vitriol, salts of zinc and charcoal are the best means for deodorizing matter under putrefaction. 3. The powders of Prof. Kittary, besides the properties they share in common with other carbolic disinfectants, deserve attention because of the isolated state of phenol in them and their contents of quicklime, which absorbs moisture—the principal condition of each kind of putrefaction—as also some part of the gases. 6. Chloride of lime and permanganate of potash quickly destroy the lower organisms in putrid liquids. 7. The disinfectants certainly retard the putrid processes in organic bodies, but their influence is only temporary. As a means of purifying air in dwellings their influence is very small, if not totally *nil*, because of the very small degree of concentration of their ingredients that can be used without injuring the health of inhabitants. 8. For uninhabited buildings the best disinfectants are nitrous acid and chlorine.

Difference of scientific opinion respecting disinfectants. It will be noticed that there are some points of difference between Dr. Baxter's conclusions and those reached by the St. Petersburg committee. In the words of the proverbial conundrum, "When doctors differ who shall decide?" The whole literature of the subject is as full of contradictions and differences as the conversation of a tree full of katydids on a summer night.

N. Y. Board of Health. In a "Memorandum on Disinfection," issued by the New York Board of Health in 1868, the following recommendations are made:

Disinfection of excrement. "To disinfect privies, water-closets, close stools, bed-pans, &c., use solution of copperas and carbolic acid.

Damp places. "To disinfect cellars, vaults, stables or any damp, offensive places, use quicklime, charcoal, copperas or carbolic acid.

Living and sleeping rooms. "For sick rooms, bedrooms and closets, cleanliness, good ventilation, quicklime and charcoal.

SANITARY CARE OF PREMISES.

"To disinfect a privy, use copperas and carbolic acid solutions, mixed as above described, to the extent of two or three pints to every cubic foot of filth treated."

These directions, though probably not in accordance with the best practice of sanitary experts at this time, are simple, practical and easily followed, and for this reason I have given them in condensed form.

In the use of disinfectants we should remember that they are at best a very uncertain dependence. In operation I can only compare them to the legal document known as a "stay of proceedings," which does not set aside the judgment of the court, but merely arrests for a time the execution of its decree to afford time for fuller inquiry or for appeal to a higher court. Like the "stay of proceedings," the disinfectant is operative for a time only. We arrest decomposition and stop a nuisance for a time by this means, but if we do nothing more, the dangerous processes we seek to avert will go on as before. In the meantime we must attack the evil by more vigorous and permanent means. The decomposing matter must be removed to some place where decomposition can go on safely; the foul drain or cesspool must be cleaned out and ventilated; the horrible privy vault must be emptied, purified and filled with clean, dry earth, and the privy transformed into an earth closet; the offensive water-closet must be put in order or abolished—in a word, we must remove filth, abolish dirt and correct all conditions of known or suspected unhealthfulness. Disinfectants will not do this. Dr. John Simon, Chief Medical Officer of the Privy Council of the Local Government Board of Great Britain, in his able memoir on "Filth Diseases and their Prevention," says with much of grace as well as force: "To chemically disinfect (in the true sense of that word) the filth of any neglected district, to follow the body and branchings of the filth with really effective chemical treatment, to thoroughly destroy or counteract it in muck heaps, and cesspools, and ash pits, and sewers, and drains, and where soaking into wells and

Margin notes: Stay of proceedings. Sanitary policing of premises. Dr. Simon. Limitation of disinfection.

where exhaling into houses, cannot, I apprehend, be proposed as physically possible, and the utmost which disinfection can do in this sense is apparently not likely to be more than in a certain class of cases to contribute something collateral and supplementary to efforts which mainly must be of the other sort. This opinion as to the very limited degree in which chemistry can prevail against arrears of uncleanliness does not at all discredit the appeals which are constantly and very properly made to chemistry for help in a quite different sphere of operation—with regard, namely, to the management of individual cases of infectious disease, and to the immediate disinfection of everything that comes from them. In this latter use of disinfectants everything turns on the accuracy and completeness with which each prescribed performance is done; but such accuracy and completeness are, of course, only to be insured when operations are within well-defined and narrow limits, and in proportion as disinfection pretends to work on indefinite quantities or in indefinite spaces it ceases to have that practical meaning. Again and again in the experience of this department a district has been found under some terrible visitation of enteric (typhoid) fever from filth infection operating through house drains or water supply, but with the local authority inactive as to the true cause of the mischief, and only bent on practicing about the place, under the name of disinfection, some futile ceremony of chemical libations or powderings. Conduct such as this, referring apparently rather to some mythical 'epidemic influence' than to the known causes of disease, and savoring rather of superstitious observance than of rational recourse to chemistry, is eminently not that by which filth diseases can be prevented, and contrasting it, therefore, with means by which that result can be secured, I would here specially note a warning against it."

Scientific use of disinfectants.

With regard to the sanitary policing of premises Dr. Simon says:

Sanitary care of buildings and lands.

"Wherever human beings are settled for residence the cleanliness which is indispensable for healthy life can only be secured by strict method. Even where houses stand singly and with wide space around them, the householder cannot safely neglect that sanitary obligation with regard to the refuse of his own household—the slop waters, cooking waste, various house sweepings, the human fæces and urine, the excrements of domestic animals, &c.; and the obligation becomes more and more important in proportion as dwellings are gathered together in comparatively small areas."

A very practical country physician was once asked by a neighbor who was not over particular as to the condition of his premises, what would be the best disinfectant to get for use before hot weather came on.

"I will give you a prescription if you will get it filled and use it," said the doctor.

A practical prescription.

This was agreed to and the doctor wrote as follows:

℞. Rake 1
Shovel 1
Wheelbarrow 1

Sig. Use vigorously every 24 hours until relieved.

The hint was taken and the premises were cleaned up. Sunshine did the rest.

CHAPTER XIII.

THE PLUMBER AND HIS WORK.

Popular abuse of plumbers. There is probably more gratuitous abuse of one kind or another lavished upon plumbers than upon all the other mechanics directly or indirectly connected with the building trades. The person who buys or leases a cheap house is not commonly surprised to find it badly built in every part. If the floors warp and shrink away from the surbases; if the doors spring, the windows stick and the moldings drop off, we shrug our shoulders and console ourselves with the reflection that, if people will build cheap houses they must expect that cheap materials and cheap workmanship will be put into them. We never think of assuming that the carpenters did not know their trades, or would not have done good work if they had been paid for it; and when repairs are needed we send for some- *Unfair criticisms.* body who can make them. When, however, we find a house badly piped, we seem to take it for granted that the plumber who did the work was an ignorant fool. As repairs become necessary, our spite against the individual plumber who piped the house by contract extends to the whole trade. We send for them under protest, growl at their bills, denounce them as frauds and swindlers, and wish we could make our home on some barren isle in mid-ocean and forever escape their miserable swindles and exactions. When the plumber whom we call in to mend the pipe in some inaccessible place has to tear up our floor or break down our walls, we never think of blaming *Blame where it does not belong.* the architect or the builder. The unfortunate plumber takes it all, and if he is not rendered unhappy by the consciousness that he is not loved by the public, it is probably because he is quite indifferent to what people think of him so long as they pay him his bills. They will come after him fast enough when his services are needed.

Popular writers who are constantly seeking targets for their wit—which consists very largely in exaggerating the vexations and annoyances of daily experiences and magnifying the commonplace—have done not a little during the past few years to make the public regard plumbers as the natural enemies of all mankind. An example of this kind of literature, which has the exceptional merit of being witty and not ill-natured, occurs in the very popular book entitled, "My Summer in a Garden," from which I quote as follows: *[The plumber's place in literature.]*

"And, speaking of a philosophical temper, there is no class of men whose society is more to be desired for this quality than that of plumbers. They are the most agreeable men I know, and the boys in the business begin to be agreeable very early. I suspect that the secret of it is that they are agreeable by the hour. In the dryest days my fountain became disabled; the pipe was stopped up. A couple of plumbers with the implements of their craft came out to view the situation. There was a good deal of difference of opinion about where the stoppage was. I found the plumbers perfectly willing to sit down and talk about it—talk by the hour. Some of their queries and remarks were exceedingly ingenious, and their general observations on other subjects were excellent in their way, and could hardly have been better if they had been made by the job. The work dragged a little, as it is apt to do by the hour. The plumbers had occasion to make me several visits. Sometimes they would find, upon arrival, that they had forgotten some indispensable tool, and one would go back to the shop a mile and a half after it, and his comrade would await his return with most exemplary patience and sit down and talk—always by the hour. I do not doubt but it is a habit to have something wanted at the shop. They seemed to be very good workmen, and always willing to stop and talk about the job or anything else when I went near them. Nor had they any of that impetuous hurry which is said to be the bane of our American civilization. To their credit be it said that I never *[Caricaturing the trade.]*

330   THE PLUMBER AND HIS WORK.

observed any of it in them. I think they have very nearly solved the problem of life. It is to work for other people, never for yourself, and get your pay by the hour. You then have no anxiety and little work. If you do things by the job you are perpetually driven; the hours are scourges. If you work by the hour you gently sail on the stream of time, which is always bearing you on to the haven of pay whether you make any effort or not. Working by the hour tends to make one moral. A plumber, working by the job, trying to unscrew a rusty, refractory nut, in a cramped position, where the tongs continually slipped off, would swear; but I never heard one of them swear, or exhibit the least impatience, at such a vexation working by the hour. How sweet the flight of time seems to his calm mind."

Reasons for popular dissatisfaction. Now in a book addressed chiefly to plumbers it would no doubt be the proper thing to say that the popular feeling of dissatisfaction with them, their bills and their work rests upon nothing more substantial than an unfounded and wholly unreasonable prejudice unworthy of an intelligent community, *Practical" plumbers.* &c. Such, however, is not the fact. It is unfortunately true that a very large proportion of those who call themselves "practical plumbers," and who set up in business for themselves, do not know their trade and are not as honest as could be desired in its practice. In my intercourse with the trade I have met many plumbers of large intelligence and much experience who exemplified in all their business relations a delicate sense of honor and an obligation not always found in bank *Good men not scarce.* parlors nor in the counting rooms of great merchants. Men of this kind are not so scarce in the trade that we need have *Second and third rate mechanics.* trouble in finding them. I have also met a great many plumbers of average skill and intelligence—honest, as the world goes, but without any very keen sense of the obligation which rests upon all men to take nothing for which they do not render a fair equivalent. I have also come in contact with a great many clumsy ignoramuses whose knowledge of the trade in which they called themselves "practical" is limited to the wiping of misshapen joints and what ideas they managed to "pick up" while carrying a bag of tools for a few months, and who would not hesitate to charge all they could get for materials and services, wholly irrespective of their value. It is the botching, the "sogering" and overcharging of such plumbers as belong to the latter class that have created in the public mind a prejudice against the whole trade. The reader, if he be a plumber, can classify himself; but it is safe to assume that he does not represent the class of which I have last spoken. Men of this stripe never read. *Botching, idling and overcharges.*

From a long and somewhat intimate acquaintance with those connected with the trade in its various branches, I have learned to regard the work of the "practical plumber" as demanding high and peculiar qualifications. In some respects it is the easiest of all trades to learn, and a man with average mechanical ability could, with application, make himself a good workman in very much less time than would be required to learn a majority of mechanical trades. This is an advantage to the apprentice, in so far as it enables him to become a good workman in a comparatively short time; but there is constant danger that the ease with which the practice of the plumbing shop may be learned will encourage laziness on the part of the apprentice and a disregard of the obligation which rests upon every mechanic to master the theory as well as the manipulations of the trade he essays to learn. The work of the plumber looks so simple to the apprentice, and is so simple in many respects, that before he has carried the tools for six weeks he imagines he knows it all, and unless he be a young man of exceptional good sense he gets through the balance of his apprenticeship as easily as possible, encouraged by the proud consciousness that he could wipe a joint as well as the boss if he only had the opportunity, and that on his twenty-first birthday he will set up as a "practical plumber" with as good a right to the title as nine-tenths of those who assume it. *Qualifications demanded for "practical" plumbing work. The trade easily learned. Plumbers' apprentices.*

This feeling of self-sufficiency, which is quite natural under the circumstances, commonly leads the apprentice to look with profound contempt upon study or solid reading. A majority of the workmen with whom he comes in contact know very little more of the theory of the business than he can "pick up" without much effort, and he is rarely called upon to perform any work during his apprenticeship which requires a knowledge greater than he possesses, or encourages him to study causes and investigate principles. Thus the golden opportunities of youth slip by unheeded; at the proper time he is graduated a full fledged journeyman, and after that he has, as the rule, little of either time or inclination for study. As a consequence we have a very large proportion of practical plumbers who are only practical, knowing simply the characteristic manipulations of their handicraft, but who are practically ignorant of its principles, except perhaps such as have been learned by experience and are imparted from generation to generation in the traditions of the shop. I do not mean to say that all plumbers enter upon the practice of their trade unprepared. Such a statement would be unfair and untrue. The thoroughness with which an apprentice learns his trade depends largely upon his own intelligence, character and habits, and upon the character of his employer. Those who are so fortunate as to be brought up in well-ordered shops, under the direction of men who know their business and believe it to be their duty to teach it in all its branches to their apprentices, have themselves to blame if they do not become skillful and thorough workmen. But all boys are not thus fortunate, and when left to "pick up" their trades, they are apt to pick up only so much as they can carry without straining their mental capacity.

I have said that in my judgment the practical plumber requires high and peculiar qualifications for the work he has to perform. Primarily he must be a man of sound good sense and possessed of a wide range of general information. He needs these qualifications for the reason that he must be to some extent a jack-of-all-trades. In jobbing there is no telling what kind of work he may be called upon to perform, and his success in jobbing depends largely upon being able to do the right thing first and do it in the easiest way. His general information must be comprehensive, including a knowledge of practical hydraulics, of arithmetic and algebra, of the principles of chemistry, and of half a dozen trades connected with or relating to house building. He must at times do and undo the work of the carpenter, the mason, the gas-fitter, the plasterer, the painter and the carpet layer. It is not always possible for these to follow him and repair the mischief he is compelled to do, and he should know how to repair it when necessary, as well as know how to avoid making unnecessary work for others. I have more than once had plumbers do a vast amount of unnecessary damage to walls, woodwork and carpets in my own house, and I can sympathize with those who find cause for complaint in the way which a great many of them apply their talent for pulling things to pieces. The skillful plumber needs to be a "handy man" with tools of all kinds, and this dexterity he can easily acquire if he have the sound good sense and general intelligence which I have placed first among his essential qualifications. *General handiness.*

He must be a man of quick perceptions and prompt in action, always ready for an emergency. He is often called upon to render services which are valuable to those who employ him in proportion to the promptness and intelligence with which they are performed. Unnecessary delays in responding to calls, tardiness in getting to work where instant action is demanded, and "fooling around" on any pretext when his work is done, will destroy any man's business reputation and leave him dependent upon chance custom. *Perception and promptness.*

He must not be afraid of himself or his work. Much of it is dirty and disagreeable; but it is useful and honorable, and should never be slighted out of consideration for his nose or his fingers. He need not fear that his dignity will suffer or *Not afraid of work.*

his character as a gentleman be called in question because he goes at his work like a man and does it as well as he knows how, whatever it may be.

Thoroughness He must be thorough. Few of those who employ him know whether his work is well done or not. He can cover up the worst kind of botching if he wants to, and generally get the same price for it that would be charged for better work by a better man. It is a matter between himself and his conscience. The consequences of his blundering or carelessness may be serious and far reaching. The few dollars he saves on a poor job may cost hundreds in damaged walls and furniture, or possibly bring sickness and death to happy households. He cannot afford to assume this moral responsibility for the sake of a present petty gain in money.

The plumber a sanitarian. He should be—and before many years must be—a sanitarian. The manner in which houses are drained is of vast, and as yet unappreciated, importance as affecting the public health. Much of the literature of this new and beneficent department of scientific investigation has a direct, practical bearing upon the work of the plumber. He must lead as well as follow the progress of reform now fairly begun. What has already been said and written has awakened no little popular interest in the subject of better and safer drainage systems than are now commonly employed, and before many years those will monopolize the cream of the business who are abreast with the progress of sanitary reform, and who are untrammeled by ignorant prejudices and narrow views. The plumber of the near future will be a man who can intelligently begin where the engineer leaves off, and bring any system of drainage which the former may carry out in part to its complete, perfect and scientific consummation.

Honesty. He must be honest. I do not mean by this that he should not be a thief, for in no trade of which I have any knowledge is the standard of honesty, as regards a sense of the difference between *meum* and *tuum*, higher than in the plumbing trade.

The plumber enters a house with almost a *carte blanche* to go where he will, and I am happy to say the confidence of the public is rarely abused by one of the craft. But honesty implies something more than a respect for the property rights of others. It implies honor between man and man, and this cannot exist where false charges are made or exaggerated items set down in bills. The man who wastes the time for which I pay is as dishonest, morally, as the man who picks my pocket. If he charges me with two hours time when the work done could have been finished in one hour, he does not deal honestly by me, and cannot claim to be an honest man though he respects the sanctity of bureau drawers and leaves my wardrobe unmolested. This is plain talk, but there is no reason why it should give offense to any one. No one will deny its truth. Not long ago some students in the School of Mines, in New York, were taking photographs for the use of one of the faculty in a room where a plumber and his assistant were at work. "Gentlemen," said the plumber, "suppose you take a picture of me reading a newspaper with one eye and watching the door with the other to see if the boss is coming, while the 'prentice potters around making believe he is doing something." Such a picture would be characteristic it must be confessed.

<small>What honesty implies.</small>
<small>Wasting time.</small>
<small>An incident.</small>

With regard to overcharges on materials there is more to be said in extenuation. If a man charges me $3 for what cost him $1, or 50 cents for what cost him 15, he may justify it to his conscience without much trouble by claiming that the buying, transporting and risk in handling are worth the difference between the value of the article and the price he asks for it. Sometimes they are, but oftener they are not. I will not discuss the question here. It is enough to repeat the old proverb, "Honesty is the best policy." Good work and fair charges for labor and materials are the prime, and indeed the only, conditions of sure, permanent and legitimate success in the plumbing business.

<small>Overcharge on materials.</small>

336   THE PLUMBER AND HIS WORK.

What the plumber needs to know. Now while the ordinary work of the plumber is simple and easily learned, as I have said, a knowledge of how to handle, cut and connect pipes does not make a man a master of the plumber's trade. There are a great many good workmen who are by no means good plumbers. This is a fact which the intelligent and ambitious apprentice should keep in mind, and not be misled by self-conceit and the pride of half-knowledge into the idea that his little experience has taught him all there is to know. The great evil of the trade is that a man can practice it without learning it. If it were not so the possession of a solder pot, ladle, cloth, shave hook, hammer, saw and a few other tools, a sign and a little practical knowledge, would not constitute so many men "practical plumbers."

It is not my intention in this chapter to waste space paying compliments. The reader has probably discovered this already. If he will follow me to the end, however, he will see that I consider the ignorant, incompetent and dishonest plumber the legitimate product of a system, and believe that with the abolition of that system he will disappear from the ranks of the trade and turn blacksmith's helper, horse-car conductor or something else better fitted to his abilities.

Plumbing work. I will now speak somewhat generally of plumbing work. In building a house there are many things which can be sacrificed to economy, but there are four things which cannot be too good. These are the foundations, the roof, the plumbing work *Essentials in house building.* and the apparatus for heating. The two essentials first mentioned are usually secured at any cost, but the economy comes in in the plumbing work and the furnace. The extent to which this curtailment of necessary expenditure is carried is often surprising. When people set out to build houses to live in they usually desire that they shall be healthful, comfortable, and as elegant in external and internal appointments as their means will permit. The carpenter, the mason, the roofer and the painter are all expected to do good work and charge a good price for it; but the plumber is required to make his bid below the cost of even second-class work, and the owner canvasses the market for the smallest and cheapest furnace he can find which can be driven to do the work expected of it. The fact that good drainage and pure air are the essential conditions of health and comfort is seldom taken into account. These are matters in which economy can be carried to any extent and Mrs. Grundy will not know it; consequently useless ornamentation is paid for while health and comfort are left to take care of themselves. We would naturally suppose that in this age of the world's progress a majority of the houses built to live in would be so arranged as to guard against all conditions known to be unhealthy; but such is not the case, and until the intelligent classes of the community realize more fully than they now do the importance of having good drainage at any cost, we shall continue to have economy practiced just where it can least be afforded. *[What is expected of the plumber. Health sacrificed to show.]*

As the rule, new work in this country is done by contract. The community are willing, under favorable conditions, to trust masons, carpenters, plasterers and painters to work by the day when good work is to be done; but for the reason already explained, a majority of house builders consider it advisable to bind the plumber under a contract. Now let us see how this system works. The plumber takes the architect's plan and specifications and makes a calculation thereon. If he be an honest plumber, with a reputation to protect and work enough of the kind he prefers to make him indifferent about getting contract jobs, he will make a bid at a price which will enable him to carry out the letter and spirit of the architect's specifications and leave him a fair, honest profit. If he does this the chances are ten to one he will not get the contract. If, on the other hand, he be a plumber with no reputation to lose and in want of business, to whom a contract is important, he will make his estimate upon a very different basis. He studies the architect's specifications to see where and how and to what extent he can take advantage of any errors or omissions and save in cost of *[Contract work. How the contract system works.]*

338   THE PLUMBER AND HIS WORK.

Loose specifications. materials. Usually there is plenty of chance for this, for a majority of architects draw their specifications of plumbing work so loosely and with so little knowledge of the practice of the trade, as to leave a liberal margin for "skinning" on the part of the plumber who does the work. If given to understand that the lowest bidder will get the contract, his sole study is to see how cheaply he can do the work, and the result of this study is a plan for doing it so that, even at the low price he puts upon it, he can make a profit. The price will probably be below what every intelligent plumber would know to be the net cost of the work called for by the specifications.

Why contract work is badly done. We will suppose the contract is awarded him and he goes to work. What is he to do? Obviously he must make the contract pay if he can, for he cannot afford to lose money for any one else's benefit. There is but one course open to him. He must resort to what the shipbuilders call "scamping," and his success in making the job pay depends upon his ability to do this successfully. He takes advantage of every error or oversight on the part of the architect; he uses the cheapest materials he can get and puts them together in the easiest way, and where he can depart from the letter of the specifications and escape detection he will do it, provided loss cannot be avoided in any other way. I know of instances in which, in place of lead pipes carried under floors, plumbers have used $\frac{3}{4}$-inch gas pipe, and the fraud could not be detected at the time without taking up the floor, which no one thought of doing. As all the pipes which showed were lead, the natural supposition was that all which did not show were lead also. I know of another case still more remarkable. The contract for the plumbing work in a row of houses built on speculation was awarded to an irresponsible man, who bid so low that none of those who competed came anywhere near him. He did the work, and while it was not well done, it was accepted and paid for. The houses were subsequently sold and people moved into them, but it was not long before they were "stunk out"—to use a forcible but

An instance of "scamping."

Houses without sewer connections.

somewhat inelegant expression familiar to plumbers. An inspection revealed the startling fact that in no case had any connection been made with the sewer. The soil pipe was carried down to the cellar and far enough underground to conceal the fact that it ended there. The drainage of the houses had been emptied into the cellar, and when the soil ceased to absorb it the smell gave warning of the nature of the evil to be remedied. The architect had taken it for granted that some sort of a connection would be made with the sewer, but it was not called for in the specifications and the plumber had not made it.

I do not propose to tell what I know of the methods by which cheap contract work is usually made to pay the plumber a profit. Those in the trade who have practiced these devices know a great deal more about them than I do; those who do not had better not learn. In a general way it can be said that the difference between the work called for in the intent and meaning of architects' specifications and that usually done by the lowest bidder under contract, is about as great as that which exists between gold and thinly gilded brass. It appears in every item of material used, in every detail of workmanship. There is certainly nothing in this to afford any occasion for surprise. We have no warrant for supposing that any man will for 50 cents furnish materials and do work to the value of $1. The less of that kind of business a man has the better off he will be. Plumbers work for profit; they are entitled to it; they should have it, and, under all but exceptional circumstances, they will manage to get it. If we cut down their prices they will cut down in the quality of materials and workmanship. They must do this or give up the business. The effect of this is to demoralize the trade, to encourage dishonesty, to discourage the introduction and employment of improved methods and appliances calculated to render our house-drainage systems safer and less liable to give rise to unhealthy conditions, and to bring business to a class of men who would get it under no other conceivable circumstances. There are a great many

How contracts are made to pay.

The plumber's alternative.

The effect upon the trade.

plumbers who make money out of cheap contracts without any compunctions of conscience—which is not to be wondered at under the circumstances; there are a great many who do this under protest and who consider the contract system utterly and unconditionally bad; there are some who will estimate on work when requested, but will always demand a fair price with the intention of doing good work if the contract is given to them; there are a fortunate few who are in a position to do business in their own way, and who will not take a contract on any terms. I know a man of this class who is almost daily called upon by gentlemen with whom he has conversations something like this:

A plumber who never "estimates."

"You have been highly recommended to me, sir, as a plumber who thoroughly understands the business, and I should like to have you do the work in my house. If you will stop in at my architect's, see the plans and give me an estimate, I will come in to-morrow and make a contract with you."

"Thank you," replies our friend the plumber, "but I don't think I care about the job."

"I want you to do it; I propose that it shall be well done; I intend to pay for it when it is done, and I don't propose that any second-class man shall do it."

"Well, sir," answers our friend, "if you want me to do the work I shall be happy to do it well and charge you only what is right and fair; but I will not give you any estimate nor will I sign a contract. I can't tell, nor can any other man, what the work will cost until it is done. If I fix a price I shall cheat you or cheat myself, and I do not propose to do either."

This is our friend's ultimatum. No persuasion can induce him to change his answer. That he has plenty of the best work, has made an honorable and extended reputation and stands at the head of the trade in the city in which he lives, is not to be wondered at. If all first-class plumbers would take the same stand, refusing to be tempted to bid for contracts or to agree to do any work for anybody at a price below what they be-

The policy of good workmen.

lieve to be enough to cover cost, contingencies and profit, they would soon monopolize all the business that is worth seeking. Incompetent and unprincipled plumbers thrive upon the mistaken idea which lingers in the public mind, that the way to get good work done cheaply is to have it done by contract. The experience of generations—centuries, even—has had little effect in exposing the fallacy of this notion. When none but second-class men will bid on an architect's specifications, the public will not be slow in recognizing the difference which exists between first and second class work. *Popular errors regarding work by contract.*

It is probable that one reason why so large a proportion of the plumbing work in this country is done by contract is found in the fact that a majority of people have an exaggerated idea of the profits of the plumbing business, as well as the low estimate of the standard of honesty in the trade already noted. With regard to profits, I have no hesitation in saying that they are usually moderate when good work is done. "There is not one plumber in a dozen who can afford to be honest," said one of the trade to me not long ago. Certainly there are very few, comparatively, who succeed in making anything more than a living. The largest percentages of profit are usually made out of people who imagine that they have made close contracts and are getting their work done cheaply. As regards honesty, I fail to see that the average standard in the plumbing business is above or below that of other trades. Contract work of all kinds is proverbially bad, and cheap contract work always was and always will be the standard of comparison for everything inferior in quality and transient in character. "The world seems to be going to rack and ruin," said a wealthy contractor to Foote when that famous wit was in his prime. "Why is it?" *Why the contract system is maintained* *Small profits.* *Contract work proverbially bad*

"I cannot imagine," answered Foote promptly, "unless it was built by contract."

The now historic joke only crystallized a bit of universal experience. Plumbers are probably as honest as other mechanics in carrying out their contracts, and no doubt they give as large a percentage of value for the money they receive as do masons or carpenters. The kind of competition to which they are subjected, however, from those in their own trade forces them to work cheaply and, as the consequence, to do cheap work.

<small>Competition in the trade.</small>

It is safe to assert that the economy practiced by housebuilders in the matter of plumbing work is always of the kind which saves at the spigot and wastes at the bung. I am informed by experts in the trade who are authorities on all matters pertaining to it, that the widest margin of saving on poor work as compared with good rarely exceeds 25 per cent. An average city house can be piped scientifically, with the best materials and in the best way, for about $1200, including all necessary fixtures. It is possible to make the work cost more, but this amount will pay for as much first-class plumbing work as is needed in most New York houses of the better class. The house could not be plumbed at all, provided the same plan were followed, for less than $900. The saving of $300 thus secured is a trifle compared with the sum upon which the annual expenses for repairs would pay interest; and when we consider the dangers and discomforts to which bad plumbing work gives rise, it is too paltry and insignificant to merit a moment's consideration. If we cannot afford to have the plumbing work in our houses well done, we had better have less of it. When we aspire to the luxuries of baths, water-closets, bedroom wash basins and similar refinements, we should first count the cost and see whether we can afford them. If we cannot afford to have the best materials and workmanship, we had better content ourselves with wash bowls and pitchers in our bedrooms and one water-closet somewhere out of doors.

<small>Mistaken economy.</small>

<small>No saving on cheap plumbing work.</small>

<small>Repairs.</small>

The contract system is an evil which cannot be easily or promptly reformed. No doubt arguments could be found in favor of it, and it must be admitted that the abuses of the system, rather than the system itself, need reforming. The probable cost of a job of plumbing, plus a reasonable allowance for

<small>Reforming the contract system.</small>

profit, can always be ascertained with approximate accuracy. Now it may safely be assumed that a man who agrees to do the work for less than this is either mistaken in his estimates or proposes to make the contract pay at any price. In either case it is not desirable that he should have it. If mistaken in his estimate he will, as the rule, save himself from loss if he can. There are some men who would carry out a contract in letter and spirit if it ruined them, but such men are exceptions in any trade, and, moreover, they do not often make the mistake of agreeing to do work for less than it is likely to cost them. If, on the other hand, the contract is taken with the intention of making it pay, there is little reason to hope that you will get more than your money's worth, though it be done for half price. The chances that you will get an honest plumber to cheat himself for your benefit are about as one to one hundred that you have reason to conclude, after the work is done, that you are the victim of your own smartness, and that the man with whom you made your shrewd bargain has far better reason to feel satisfied than you have. The great danger of the contract system is the temptation it offers to give our work to the lowest bidder. Plumbers who can be trusted and whose bond is good for anything, do not make any haphazard estimates. If an architect's specifications are specific, they can tell to a dollar the cost of every ounce of material called for, and with approximate accuracy, at least, the time it will take to put these materials together. To the cost every honest practical plumber will add the percentage he has learned by experience to allow for waste and contingencies, and to the sum of these a fair and legitimate profit. If we are willing to contract with him to do our work on this basis, well and good. We know in advance just what the work will cost us, and we shall probably have it well done whether the plumber's profit be a little more or a little less than he expected. But on no other basis can we afford to contract with any man for anything. Bids under the price named by a responsible plumber of character and experi-

Why the lowest bidders should not have contracts.

ence who is willing to give you a memorandum of items, can safely be regarded with suspicion. Obviously, therefore, contracts—if awarded at all for plumbing work—cannot always be given to the lowest bidder. This is a proposition so plain that no man with the average allowance of common sense can fail to see its wisdom.

Architects' specifications For much of the looseness which has crept into the morals and practices of the plumbing trade, the architects are responsible. A very large proportion of their plans and specifications are prepared with so little knowledge of the principles of plumbing work that it would be impossible to pipe a house in accordance with them. The plumber cannot be held responsible for their errors or mistakes, but for his own protection he is very apt to take advantage of them. As I have spoken of this subject in another chapter, I shall not discuss it here.

How the evils affecting the trade may be reformed For the evils of ignorant and dishonest plumbing there is but one remedy which promises to be permanent and certain. It is to employ only skillful and honest men who will not agree to work for less than fair prices. When this is provided for in advance it makes, practically, but little difference whether our *The percentage system.* work is done by contract or for a percentage. The latter system has many advantages, however. The plumber who works for a percentage, usually ranging from seven to ten, according to the size of the job, agrees to bill materials and labor at their net cost and take the percentage of the total agreed upon as his profit, including the superintendence, &c. If the builder or house owner prefers, he can buy his own materials and the plumber will furnish the labor required to put them together. *Its advantages.* This insures good materials and good workmanship, and costs no more than any man should be willing to pay for work he has done. The fact that a majority of our best plumbers are willing to work on this system, shows that they are content with fair profits and ready to give their customers every *Jobbing.* reasonable advantage. Where job work is to be done, such as repairs and alterations, the customer has but one means of protecting himself. He must intrust his work to some man who has a reputation for honesty and fair dealing. The moment he begins to haggle about the price of work before it is done, he invites the plumber to cheat him in order to save himself. In a word, it is with the plumbing business as with all other trades—if you want good work and fair dealing you must deal with good men and pay fair prices. *Good pay for good work.*

How do the plumbers regard this subject? I believe that a majority of those who will see this book will agree perfectly with everything I have said in this chapter. Perhaps I cannot furnish better proof of this than by quoting from a few of the many letters I have saved out of an extensive and interesting correspondence with representative men in the trade, extending over a period of several years. These letters are *bona fide*, and my quotations are given verbatim. *The views of the trade.*

A plumber of thirty years' experience, doing business in Syracuse, N. Y., sent me a letter, called out by a published article of mine, from which I quote as follows: *Extracts from correspondence.*

"You say that the responsibility which rests upon the plumber is often more serious than he imagines, and that ignorance is, at best, a poor excuse for the mischief which may result from his mistakes. *A plumber's responsibility.*

"Now, I admit that this trade, like most others, is imperfectly learned in America, because we have no apprentice system worthy of the name; but the worst feature of the case is that builders and owners of houses think they know as much about plumbing as the man who has served a lifetime at it. It is this dangerous ignorance, mistaken for knowledge, which enables employers who know but little of the business and who hire cheap men, to get contracts for plumbing work, because they will follow the directions laid down for them by men who know still less than they do. If you will make inquiry in the trade I think you will find that about one-half of the so-called 'practical plumbers' cannot lay out a job so that it will work right when finished. I call to mind laborers, masons, hardware *The trade imperfectly learned.* *Practical plumbers.*

dealers, jewelers, carpenters, tinmen, machinists, a county sheriff and a tanner, who think they know all about plumbing and can make money out of it. A member of a prominent house in your city, dealers in plumbers' supplies, told me only a few weeks ago that they had lately received an application from a man for their catalogues, with list prices, &c., and *all the information they could give him.* He knew nothing of the business, but was going to set up a shop, as several men in his place had done very well at it, and he believed there was money to be made in the business. It is such men as this who do the kind of work you justly characterize as 'unscientific plumbing.'"

Another correspondent, a successful and well-known plumber in Boston, comments as follows on some views expressed by me in a paper read before the Public Health Association and subsequently published in the *Sanitarian:*

Incompetent architects

"As to faulty plumbing work, in most part it lies with incompetent architects and very often with gents of that profession who think themselves well posted. They of course can design the plans of a house, locate where the plumbing work is to be, write a very elaborate specifications, &c., get half a dozen plumbers to estimate with a knowledge from the start who is going to do the job. They will call in the specification for certain places, 'AAA pipe;' for other places, 'AA pipe.' At the same time they don't know one from the other except they see the trade-mark. They will come into a building; they see the ends of pipes sticking out where shown on the plan, and

How they supervise work.

see the trap for a closet put in. They simply take a bird's-eye view of it and pass on; may possibly sing out, 'Plumber, are you sure that is right?' They know no more how it is put in than a school-boy, for they do not examine it. Then again, sir, I confess there is a good deal of the fault with the plumber. The plumber is the architect's man—that is understood. The plumber is the man of that worst of leeches, the house agent. The house agent and the architect know each other. The plumber, between the two, is in a sweat box. I am giving you these plain, simple facts before going into details of the causes of defective plumbing work or advancing an idea for a remedy. Now the plumber, having to give 10 per cent. to one and 10 per cent. to another, must curtail from the AAA pipe and the AA pipe in specification; and where cast-iron soil pipe is called for, calked with molten lead, I will guarantee that more than two-thirds of the hub is filled with paper and sand. It won't leak water—oh, no—because the end of the pipe is let into the hub and has generally a run. But will it leak sewer gas? Oh, yes—because there is not lead enough there to keep it back; and all this is done under the eye of the experienced architect. Then again, sir, a great deal of the fault is with a contemptible set of house-building speculators, who probably do not own $200 in the whole block when the buildings are started. What do they care how the plumbing or any other work is done if they can make a few thousands, honestly or not? *How the plumber is bled. Making cheap work pay. The consequences. House building speculators.*

"Then again, not a little rests with penurious (honest pay, though) house owners, who are always trying to make something a little less answer the purpose. *Penurious house owners*

"The class of men styling themselves carpenters and builders are also responsible for much of the cheap and inferior contract plumbing work. For example, a person contemplating the building of a house—worth, say, $6000—goes to an architect and gets a set of plans and specifications, which of course includes the work of the plumber, roofer, painter, &c. The carpenter and builder estimates on the whole job, without consulting any of the mechanics upon whom he must rely in carrying out the plans. We will suppose that he gets the contract. He goes to the plumber, with whom he has acquaintance, and says: 'John, I have the contract to build a house for Mr. ——, at such a place, and I want you to estimate on the plumbing; but remember, I want it done as cheap as you can do it. I have taken the job so low that it is only to keep my hands going.' The same story is told in confidence to half a dozen plumbers, *Carpenters and builders. How low bids are secured.*

and the consequence is that a very imperfect job is done, and perhaps without so much profit to the plumber as would represent the price of a potful of solder. There are, of course, a great many plumbers who will not do work on this basis; but there are, unfortunately, a great many who will take anything, and so long as these can get work to do for parsimonious house-builders, so long will we have bad plumbing with its attendant evils.

<small>Competition among plumbers.</small> "I must confess there is a good deal of the fault with plumbers in trying to cut one another out of work until there is not a scrap of solder profit left. In my opinion, with the foregoing facts, it is impossible to have other than defective plumbing work."

This, it should be remembered, gives the experience of a plumber, and is not to be classed with the generalizing of one not practically acquainted with the business. Another plumber, long established in business in New York, gives us an insight into the kind of competition which those in the trade experience from their fellow-craftsmen. I quote as follows:

<small>A divided responsibility for bad work.</small> "Allow me to give you a view into the plumbing trade and how the plumbers act toward each other, and how it is that they are mainly responsible for the bad estimation in which they are held by the public, and why house owners are responsible for defective plumbing, sewer pipes, &c. Most people think that plumbers, as a trade or body, are more leagued together and more loyal to each other and the trade than any other class of workmen in New York city, but, with a few exceptional cases, quite the contrary is the fact. When it is possible to cut one another out of custom they are bound to do it. Now to come to the point, suppose you have plumbing work in your house. A pipe bursts, you send for your plumber, but as neither he nor his men are in you send around the corner for some one else. That some one comes; he takes a view of matters, and then does what you should have done—shuts the water off. Then, instead of at once beginning the job, he will begin find-

ing fault with all the plumbing work in the house, until he makes you believe that he is the only workman in New York that knows anything, and that the man who has been doing your work for years is nothing but a fool. He makes you believe also that if your pipes were altered thus and so there would be no chance for any more bursting. The upshot of the matter is you keep sending for the smart man until there is something disturbed, perhaps under your floor or it may be in the cellar, that your smart man knows nothing about, and the first intimation you have of the matter is your need for a doctor or perhaps the undertaker. And all this happens because you did not shut off the water and have patience until the fool came home to get your order. People, I know, are not to be blamed, in case a pipe bursts, for sending for the nearest plumber; their haste is usually the result of ignorance as to the danger which menaces the boiler or some other part of the plumbing work. All that is necessary in case a burst occurs is to shut off the water, and every adult member of every household should know how to do this. Then open the hot-water cock in the kitchen sink, or over the bath, keep a moderate fire in the range, and the boiler will last for 48 hours without harm. The water in it will bubble and boil, but that is all. Your plumber, when he comes, will know how to handle matters.

Unfair competition.

How plumbers profit by popular ignorance.

"Again, house owners are largely responsible for defective work, for the reason that they often impose upon the plumber the disagreeable necessity of working for the cook or the coachman. He must please these potent officials of the household. If not, they complain, and on the strength of their complaints the plumber is dismissed and a new man is employed who will do things as the cook or the coachman may be pleased to direct, and who will also 'make it right with them.' As the rule, the new man undoes or spoils the work of the old. By neglecting to exercise a judicious oversight in such matters, the householder has himself to blame for the defective sanitary condition of his

Working for servants.

The duty of householders.

premises. I would, in conclusion, advise all householders to find out for themselves whether their plumber is an honest man and does his work properly. If so, he would do well to disregard all complaints of servants and interested parties. When you find you have a good man, trust him as you do your family physician. The comparison is not a bad one, for to his skill, intelligence and judgment you owe immunity from a large variety of causes of disease. Good plumbing is of vastly more consequence than all your rich furniture and costly carpets."

I might fill many pages with similar quotations from my correspondence, but those already given will show how plumbers regard the evils affecting the trade. They are the almost confidential utterances of men who speak from long experience, and every plumber among my readers will agree with them as well as with my own general remarks on the subject.

Advice to young men in the trade. I will conclude this chapter with a few words of friendly counsel to young men in the business. There is in a great many trades a feeling that the more repairs there are the better it will be for the prosperity of the trade. A hard winter that bursts pipes in all directions is very generally regarded as a good thing for plumbers and welcomed accordingly. Mechanics often say: "We like to have work wear out because it gives us jobs." This feeling is the cause of a great deal of willfully poor plumbing work. Now, this is killing the trade. Poor workmen find employment, the business is demoralized and good men have altogether too little to do. So bad is plumbing work in general that people have just as little of it as possible. They know that in a few years the repairs will more than equal the first cost, and in putting up a new building they have, as the *The workman's true aim.* rule, only the few indispensable fixtures. The true aim of the workman, therefore, should be to make the work good and durable. If it were certain no repairs would be necessary on the plumbing of a house for a long series of years, there would be much more work put into a house. The increase of new work would more than equal the amount lost in repairs, while the quality of the work and the character of the workmen must necessarily be much higher than now.

When an accident happens which makes large repairs neces- sary, or when there is a general freeze-up of pipes, every individual in the community is injured to a certain extent. There is so much value wasted and the community is poorer. It is a short-sighted selfishness that regards such misfortunes as blessings to the trade. The wise policy is to make each job thorough, so that repairs may not be needed. Repairs

Some years ago I overheard a conversation between two plumbers as to the best methods of doing work of a certain kind. Two kinds of workmen.

Said one: "I don't want to do my work too well, or I shall have no repairs to do."

Said the other: "It has always been my policy to do work so well that it won't need repairs, and the consequence is I am always full of good work while you are always jobbing."

To the young plumber, ambitious of honorable success in life, I offer the following brief and easily remembered advice: Precepts orth remembering.

Learn your trade thoroughly.

Study its literature and learn all you can of every subject which bears upon it, directly or indirectly.

Do no work that you cannot point to with pride.

Make no charge which you could not, upon oath and with a clear conscience, declare to be, in your judgment, just and reasonable.

Deal honestly and honorably with all men, and do unto others in the trade as you would have them do unto you.

These may seems like platitudes, and so they are; but they are the sole conditions of honorable and permanent success. The man who follows this policy through life, and who combines industry and thrift with right principles, will merit success whether he achieves it or not.

Plate I.
PLAN OF FLOORS OF A CITY HOUSE,
Showing Position of Fixtures.
(See Page 120.)

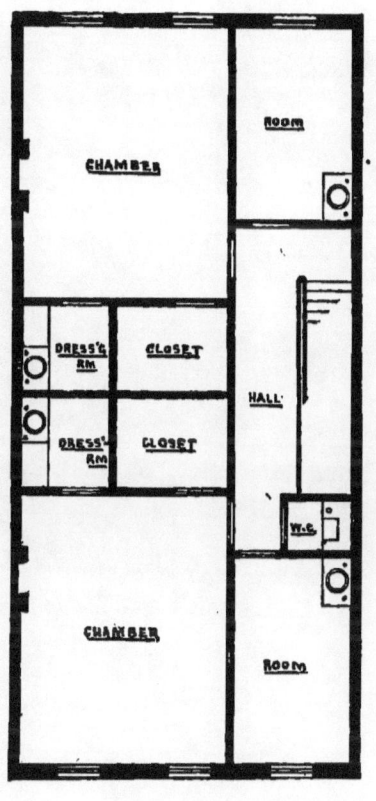
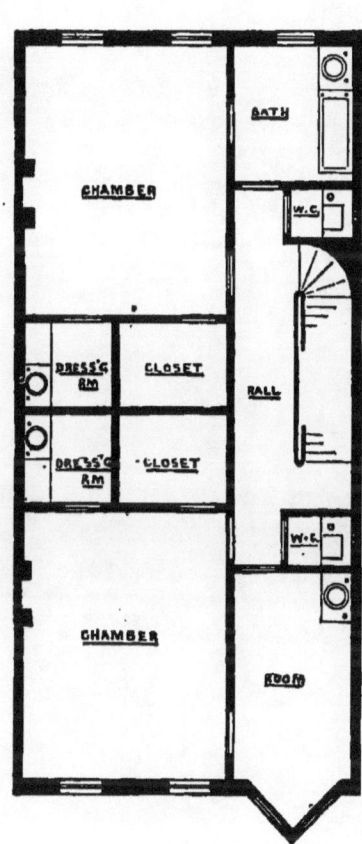

INDEX.

	Page.
Absorption of Gases by Water	93
Acids, Action of, on Lead	157, 165
With Bases, Combination of	166
With Bases, Unstable Combinations of	172
Advice to Young Plumbers	350
Ærial Disinfection	322
Air Chambers	227, 228
Chambers, Continuous	134
Cocks	117
In Water	152
In Water Pipes, Illustration of Action of	231
Traps in Water Pipes	230
Alkalies in Water, Action of, upon Lead	160
Amateur Water Analysis	298
American Houses	11
Ammoniacal Compounds in Sewer Gas	26
Amsterdam, Lead Poisoning in	180
Analyses, Limitations of Amateur	197
Analysis, Concentration of Water for	197
Filtration of Liquid for	198
Ancient Wells, Remarkable	299
Animalculæ, Destruction of	310
In Water, Vitality of	309
Anti-Infectants	316
Antiseptics	317, 321
Apprentices in the Plumbing Trade	333
Architects, Incompetent	346
Mistakes of	16
Specifications of Plumbing Work	344
Architectural Practice, Evils in	12
Architecture, Conservatism in	21
Hygiene in	11
Artesian Wells	162, 303
For Town and Village Supply	304
Ash Barrels	313
Atmospheric Contamination by Sewer Gas	75
Pressure, Hight to which Water is Raised by	237
Back-Door Nuisances	275
Bad Air	14
Baker, Prof., on Disinfectants	314
Barns and Barn-yards	263
Baxter's, Dr., Report on Disinfectants	321
Beale's Theory of Fever Contagion	34
Belgrand, Mons	148
Bichromate of Potassa Test for Lead, the	199
Bidders, Lowest	343
Black Assize, the (1577)	9
Blind Drains	279
Blowing through Seals in Branch Wastes	72
Bobierre's Experiments with Lead	154
Boiled Water, Restoring the Flavor of	310
Boiler Connections	123
Boilers, Accidents to	125
Device for Cleansing	126
Explosion of Kitchen	125
Kitchen	125
Vacuum and Safety Valves for	125
Waste Cocks for	126
Water from	126
Boiling Water, to Purify it	310
Borates	165
Boxing Pipes	131
Braces for Pipes in Wells	242
Brass Service Pipes	111
Bromides	178
Buchanan, Dr., on Earth Closets	270
Buckler's, Dr., Experiments	190
Buel, R. H	231

	Page.
Bursting of Pipes, Means of Preventing the	134
Reason for the	127
Calking	47
Pipes in Niches, Difficulties of	51
Capacity of Pipes, Relation of Length and Diameter to	118
Carbolic Acid as a Disinfectant	317, 322
Carbonate of Lead	159
Of Lime as a Protection to Lead	169
Of Magnesia and Iron	170
Of Soda	170
Carbonates	165
And Sulphates in Waters Supplied to Cities	172
Formed on Lead, Causes which Defeat the Protective Action of	171
Carbonic Acid	165
Gas	171
Iron Dissolved by	206
In Sewer Gas	24
In Water	297
Carpenters and Builders	347
Carrying Service Pipes into Houses	131
Cassamajor's Experiments	189
Cast-iron Waste Pipes	45
Cellars, Causes of Sickness found in	261
Sanitary Care of	314
Under Country Houses	261
Wet	262
Cement Joints	49
Cement-lined Tanks	140
Cerebro-Spinal Meningitis in New York	238
Cesspool Drainage	275
Pumps	276
Cesspools, Backflow from	278
Construction and Care of	276
Gases of	24
Ventilators for	277
Chain Pumps	239
Charcoal as a Disinfectant	320, 321
Filters	308
In Cesspool Ventilators	273
Chemical Action of Sewer Gas	28
Composition of Sewer Gas	24
Tests for Tin	212
Chemistry of Plumbing, the	147
Chester in the Sixteenth Century, Mortality in	8
Chloride of Lime	319
Chlorides	163
Action of, on Lead	175, 176
Formation of	175
In Sea Water	176
Solubility of	175
Chlorine Gas, Disinfectant Powers of	320
Chlorine in Potable Water, Tests for	293
In Water	294
Cholera in London	31
Christison's Experiments with Lead	152
Circulating Pipes	
Circulation	121, 123
Cistern Pumps	239
Durability of	240
Cistern Safes and Overflows	145
Cisterns	229
Capacity of	145
Elevated	144
For Rain Water	144
Underground	144
Cities, Neglect of Health in	312
Citric Acid	310

INDEX.

	Page.
City Houses, Characteristic Smell of	97
Cleanliness and Dirt	312
Coal Tar Products	318
Cochituate Water, Carbonates and Sulphates in	172
Cocks and Faucets	126
Combination of Acids with Bases	166
Competition among Plumbers	342
Composting Fæces	267
Composting, Theory and Methods of	264
Concentration of Water for Analysis	197
Conditions of Health	323
Connecting Boilers with Water-Backs	125
Conservatism in Architecture	21
Constant Service, Importance of a	104
Constipation Induced by a Lack of Suitable Privy Accommodation	268
Contagium, Generation of	323
Contents of Pipes	218
Contract System, How it Works in the Plumbing Trade	337, 339
Reforming the	342
Contract Work	337, 341
Copper, Action of Water on	213
And Brass in Contact with Lead	189
And Other Metals, Galvanic Action between	213
Kitchen Utensils, Danger of	214
Salts of	214
Sewer Pipes	111
Corrosion in New and Old Lead Pipes, Activity of	192
Of Lead Pipes by Sewer Pipes	74, 75
Of Pipes	106
Of Ship Plates by Galvanic Action	188
Corrosive Salts, Mixtures of	184
Country Districts, Causes of Unhealthfulness in	288
Neglect of Health Precautions in	289
Water Supply in	291
Country Houses, Sanitary Construction and Drainage of	258
Craven's Tests of Pipes	225
Creeping of Lead	112
Croton Water	162
Carbonates and Sulphates in	172
Sediment, Experiments with	296
Croydon, Sanitary Works of	38
Typhoid Fever in	30
Curbing for Wells	303
Dana's, Dr., Opinion of Lead Poisoning	202
Decay, Gases of	28
Decaying Organic Matter	312
Vegetable Matter	33
Decomposition in Cellars	262
Of Organic Matter	315
Defects in Pipes	191
Defective Joints in Waste Pipes	47
Defective Trapping	67
Deodorants	317
Disease, Causes of	6
Communicated by Sewer Gas	35
Diseases Conveyed by Germs	27
Disinfectants	267, 316
Scientific Use of	326
Disinfection	314
By Heat	323
Differences of Scientific Opinion Concerning	324
Limitations of	325
New York Health Board's Memorial on	324
Of Damp Places	324
Of Excrement	324
Of Foul Matter in Masses	322
Of Foul Waste Pipes	78
Of Living and Sleeping Rooms	324
Of Water-Closets	93
Russian Report on	323
Dip of Traps	80, 81
Distribution of Water in Houses	115, 120
Drainage, Dangers of Cheap	62

	Page.
Drainage, Defects in House	19
Importance of Good	44
Of Country Houses	274
Of European Cities	43
Of Lands	259
Of Roofs	229
Drains for Country Houses	275
Draught upon Traps	66
Driven Wells	305
Drive-Well Tubes	305
Droughts	146
Dry Conservancy Systems	86
Dumas' Experiments	170
Earth as a Disinfectant	270
Closets	266, 269
English Origin of	270
Price of	272
Commodes, Home-made	272
Privy, How to Make an	266
Sewage System, the	283
Economy of Power in Pumping	236
Electrical Relations of Metals	168
Empiricism and Superstition in Medicine	5
Emptying Pipes	112
Encasing Buried Pipes in Larger Ones	131
Epidemics in Country Districts	290
In Small Towns	301
Mediæval and Modern	7
Prevailing	36
Essentials in House-building	336
Estimates, Plumbers'	340
Europe in the Middle Ages, Life in	7
Evaporation of Seals	76
Excrement, Disinfection of	324
Expanding Alloys	48
Expansion and Contraction of Iron Pipes	51
Of Metal Service Pipes	112
Of Air in Sewers by Heat	65
Explosion of Kitchen Boilers	125
Extracts from Plumbers' Correspondence	345
Fergus, Dr., Experiments with Lead Pipes and Traps	74, 77
Ferrules	52
Substitutes for	52
Fever and Ague	259
Fever Contagion, Beale's Theory of	34
Germs	34
Nests	6
Fields' Flush Tank	280
Filled Lands	258
Filling Joints with Lead	48
Filter Pipes	198
Tank for Kitchen Drainage	286
Filtering Mediums	137
Filters, Charcoal	308
Cleansing of	136
For Water Containing Lead	194
Foul	136
Iron Sponge	309
Magnetic Iron	195
Filth Poisoning	18
Filtration of Liquid for Analysis	198
Of Water	135
Through Earth	284
Flow of Small Streams	129
Fluids, Presence of	217
Flush Valves	99
Flushing Closets by Direct Connections with Service Pipes, Dangers of	100
Flushing of Traps	76
Sewers	38
Fordos' Experiments	172
Freezing in Pipes, Precautions Against	132
Of Underground Pipes	130
Of Water in Glass-lined Pipes	110
Frictional Resistance of Small Pipes	118
Friction of Pipes	210

INDEX.

	Page.
Frozen Pipes, Evils Resulting from	128
Methods of Thawing	132
Frost, Protecting Pipes from	58
Protecting Service Pipes from	127
Furnace Heat, Healthfulness of	13
Furnaces, Hot-Air	13
Galvanic Action between Copper and Other Metals	213
Between Zinc and Iron	209
Corrosion of Ship Plates by	188
Lead Corrosion by	187
Galvanized Iron, Cassell's Experiments with	208
Pipes	108
Tanks, Water from	208
Garbage Bins	313
Gases Absorbed by Water	99
Of Decay	28
Passed through Water	77
Gaskets	
Generation of Contagium	323
Germs, Diseases Conveyed by	27
Fever	34
Germ Theory, Chemical Aspects of the	31
Explanation of Sewer Gas Poisoning by the Of Disease	36
	26
Glasgow, Typhoid Fever in	74
Glass-lined Water Pipes	109
Glass Service Pipes, Manipulation of	110
Gaol Fever	8
Grease Traps	45
Hallock's Experiments with Tin in Saline Solutions	211
Handiness with Tools	333
Hand Pumping	236
Hardness of Water, Determination of	292
Permanent	292
Head of Water, Calculation of	217
Consumed by Friction of Pipes	220
Rule for Finding	221
Head, Pressure of Water Due to	104
Health, Conditions of	323
In Cities, Conditions Affecting	43
Sacrificed to Show in Architecture	337
Heat, Disinfection by	323
Honesty among Plumbers	334
Hopper Closets	94
Horsford's, Prof., Experiments	185
Hot-Air Furnaces	13
Hot Water Distribution	121
Service	122
House-Building, Essentials in	336
In the United States	11
Speculators	347
House Connections with Cesspools	276
Drains of Stone	60
Drains, Tile	60
Drains, Wooden	60
Houses without Sewer Connections	333
Hugo, Victor, on the Sewage Waste of Paris	86
Hydraulic Rams	353
Hydraulics Applicable to Plumbing	216
Hydrochloric Acid	310
Hygiene in Architecture	11
In its Practical Relations to Health	5
Impurities in Spring Water	306
In Water	136
Indoor Commodes, Sanitary Importance of	269
Infections and Contagions	316
Infusoria	31
Inorganic Poisons	28
Intermittent Downward Filtration System, the Water Service in Houses	279
	116
Iodide of Potassium Test for Lead, the	200
Iodides	178
Iron and Lead Connections	52
Iron as a Material for Service Pipes	204

	Page.
Iron, Chemical Action of Water on	205
Dissolved by Carbonic Acid	206
Influence of Salts in Water on	206
In Water	108
In Water, Tests for	207
Mains	148
Pipes, Conditions of Safety in Use of	46
Expansion and Contraction of	51
Joints of Lead and Iron	185
Sizes and Weights of	226
Supports for	113
Supports for Vertical Lines of	51
Thawing Ice in	133
With Tin Lining	109
Wrought	107
Piping in Houses	115
Protective Oxidation of	206
Pumps	239
For Outdoor Work	241
Rust, Corrosion of Lead by	186
Sponge Filters	309
Sulphide and Sulphate of	173
Tanks	140
Waste Pipes, Weight of	58
Jacketing Pipes	129, 131
Jews, Sanitary Laws of the	23
Jobbing	344
Joints, Cheap Filling for	50
In Iron Pipes	47
In Iron Pipes Made with Putty	49
In Tin-lined Lead Pipes, Corrosion at	190
Made with Cement	49
Made with Expanding Alloys	48
Made with red Lead	48
Made with Rubber Washers	49
Rust	50
Sulphur and Pitch	50
Wiped	106
Kitchen Drainage, Filter Tank for	285
Refuse	287
Laboratory Experiments, Sources of Error in	186
Land Drainage	259
La Pierre's Experiments with Lead	151
Lateral Pressure of Fluids	217
Leaching Cesspools	275
Lead, Action of Air and Water on	153
Action of Acids on	157, 165
Alkalies in Water upon	160
Carbonate of Soda on	170
Carbonic Acid Gas on	171
Chlorides on	175, 176
Iodides and Bromides on	178
Lime on	160
Mineral Salts on	151
Mixed Salts upon	168, 181, 182, 184
Moisture and Carbonic Acid on	154
Nitrates and Nitrites on	174
Organic Matter on	178, 184
Physical Influences on	191
Potash and Soda on	159
Pure Water on	151
Sulphates on	172
Sulphide and Sulphate of Iron upon	173
Sulphur Waters on	173
Vegetable Acids on	178
An Accumulative Poison	201
And Iron Pipes, Joints of	185
And Tin, Electrical Conditions Induced by Contact of	193
As a Plumber's Material	147
Bicromate of Potassa Test for	193
Carbonate of	159
Carbonate of Lime as a Protection to	169
Causes which Defeat the Protective Action of Carbonates Formed on	171
Chemical Composition of	150

INDEX.

Lead, Chemical Law of the Action of Salts upon. 164
  Copper and Brass in Contact with. 189
  Corrosion by Galvanic Action. 187, 190
    By Iron Rust. 186
    By Well Water. 163, 164
    From Contact with Decaying Wood. 180
  Corrosive action of Water on. 167
  Creeping of. 112
  Effect of Heat in Promoting Corrosion of. 155
  Faucets for Vinegar Barrels. 157
  Filling Joints with. 48
  Filters for Water Containing. 194
  Impurities in Commercial. 150
  Infrequent Corrosion of. 162
  In Connection with Solder, Corrosion of. 171
  In Liquors used as Beverages. 158
  In Pastry and Confectionery. 200
  In Spirituous Liquors. 200
  In Water, Action of Carbonic Acid on. 200
    Amount of. 193
    By Analysis, Detection of. 196
    Standing in Pipes. 155
  Joints in Iron Pipes. 47
  Limited Power of Protective Salts Formed on. 168
  M. Besnon on the Corrosion of. 150
  Phosphate of. 173
  Physical Properties of. 150
  Pipes. 105
    Activity of Corrosion in New and Old. 192
    Corroded by Sewer Gas. 74
    Decay of Organic Matter in. 179
    Duration of Corrosive Action in. 192
    Efforts of Sharp Bends in Inducing Corrosion of. 191
    In Ancient Cities. 148
    In France. 148
    In Massachusetts, Investigations Concerning. 163
    In Mortar, Corrosion of. 160
    In Soda Water Fountains. 171
    Insoluble Coatings for. 195, 196
    Opinions of Ancient Authorities Concerning. 149
    Protection against Corrosion in. 194
    Stretch of. 113
    Unsafe. 19
  Poisoning at Tunbridge. 183
    By Snuff. 200
    By Water. 193
    Due to Chlorides, Evidences of. 176
    English Scientific Commission's Report on. 147
    Expedients for Protection Against. 194
    In Amsterdam. 180
    In Manchester, Eng. 156
    In Massachusetts. 147
    In New York. 155
    In Salem. 168
    Occupations of Victims of. 201
    Susceptibility to. 149, 203
    Symptoms and Characteristics of. 204
    Tanquerel's Observations of. 200
    Various Causes of. 200
  Quantity Required to Exert a Poisonous Influence. 203
  Roofs. 113
  Salts of. 166
  Salts, Solubility of. 167
  Shrinkage of, in Cooling. 47
  Sulphide of Ammonium Test for. 199
  Sulphide of Potassium Test for. 199
  Sulphuretted Hydrogen Test for. 198
  Summary of Facts Concerning the Action of Potable Waters on. 193
  Tanks. 141
  The Iodide of Potassium Test for. 200
  The Sulphuric Acid Test for. 200
  Waste Pipes. 44

Letterby's, Dr., Report on London Wells. 298
Leverage of Pump Handles. 236
Lewes, Typhoid Fever in. 100
Lift Pumps. 241
Lime and Lead, Experiments with. 1-7
Lime, Lead Corroded by. 160
Liverpool Water, Carbonates and Sulphates in. 172
Loch Katrine, Water of. 161
London, Cholera in. 31
  In the Twelfth Century. 7
  Plagues, the. 9
  Water, Carbonates and Sulphates in. 172
  Water Supply. 184

Magnetic Iron Ore Filters. 195
Main Wastes and Branches. 58
Malaria. 32, 301
Manholes. 65
Manure. 264
Marais' Experiments. 171
Massachusetts Institute of Technology, Tests for Lead in Water at the. 156
Materials for Service Pipes. 105
Meaning of Chemical Names ending in "ate" and "ide". 165
Medical Profession in Sanitary Work. 5
Medicine, Empiricism and Superstition in. 5
Megara, the. 288
Metallic Salts in Water. 297
Metals, Electrical Relations of. 188
  Positive and Negative. 288
Middle Ages, Life in Europe in the. 7
Mineral Impurities in Spring Water. 306
Mistaken Economy of Cheap Plumbing. 342
Modern Conveniences. 11
Morning Mists on Wet Lands. 259
Motive Power for Pumps. 235
Moule, Rev. Henry. 270
Muir's Experiments. 174, 181
Mushroom Strainers. 242

Nessler Test, the. 294
Nevins, Dr. 167
New York, Deaths from Zymotic Diseases in, in 1866-1876. 17
  Lead Poisoning in. 155
Nichols', Prof., Experiments with Zinc. 210
Nitrates. 163
  And Nitrites, Action of, on Lead. 174
  Tests for. 293
  In Water, Sources of. 174
Nitrites. 165
Nitrogen in Sewer Gas. 25

Obstructions in City Mains. 232
  In Pipes. 230
Ocean Salts in Rain. 161
Organic Acids, Character of. 159
  Decay in the Dark. 31
  Gases. 32
  Germs. 27
    Destruction of. 24
  Impurities in Water. 159
    In Well Water. 300
  Matter, Decomposition of. 315
    In Combination with Other Substances. 184
    In Waste Water. 274
    In Water, Action of, on Lead. 179, 180
    In Water Supplied to Cities. 296
    In Water, Tests for. 294
  Vapor in Sewer Gas. 26
Organisms in Air. 27
Overcharges on Plumbers' Materials. 335
Overflow and Safe Wastes, Automatic Flushing of. 56
Overflows. 54
  Wastes from. 55
  Water-Closet. 55
Oxygen in Water, Dissolved. 297

INDEX.

	Page.
Pan Closets	90
Defects of	91
Improvement in	92
Poisonous Gases from	91
Venting Receivers of	92
Pan-Closet Receivers	90
Paris in the Twelfth Century, Streets of	7
The Sewage Waste of	86
Percentage System, the	344
Permanganate of Potash	309, 319, 321
Petroleum in Sewers	42
Phosphate of Lead	173
Phosphates	165
Phosphoric Acid	165
Pig-Styes	263
Pipe Connections for Pumps, Proper Sizes of	240
Hooks	51
Pipes, Accessibility of Service	114
Actual Discharge of	221
Brass and Copper	111
Capacities of	121
Circulating	124
Composition	111
Conditions of Galvanic Action in	187
Contents of	218
Corrosion of	106
Craven's Tests of	225
Defects in	191
Decay of Organic Matter in	179
Difficulty of Thawing Buried	130
Discharge of Small	222
Discharge of Water from	218
Encasing Buried	131
Fractured by Settling of Walls	62
Freezing of Buried	130
Friction in Small	118
Friction of	219
Galvanized-Iron Service	108
Glass-Lined	109
Importance of Free Access to	57
In Deep Wells, Braces for	242
In Niches	57
Insoluble Coatings for Lead	195
Jacketing	129
Joints of Lead on Iron	185
Lead	105
Margin of Safety in	223
Materials for Service	105
Non-Metallic Substitutes for Lead	105
Obstructions in	230
Protection from Frost	58
Protection in Cold Weather for Service	127
Relative Capacity of Large and Small	119
Rule for Finding Diameter of	221
Discharge of	221
Length of	221
Sizes and Weights of Wrought	226
Strains on	191
Strength of	223
Block-Tin	225
Wrought-Iron	227
Supporting Iron	113
Tin	105
Tin-Lined Lead	106
Tinned Iron	109
Weight and Strength of Tin-Lined Lead	225
Why They Burst	127
Wrought-Iron Service	107
Zinc as a Material for Service	207
Piping with Iron	115
Plagues, the London	9
Plates 1 and 2, Description of	120
Playfair, Dr. L., on Cesspools	276
Plumber and His Work, the	328
A Sanitarian, the	334
Qualifications of the Practical	331
What is Expected of the	337
Plumbers, Advice to Young	350
Apprentices	331

	Page.
Plumbers, Honesty Among	334
Popular Abuse of	328
Responsibility, the	345
Plumbing Fixtures in Houses	18
Practice, Examples of Bad	53
The Chemistry of	147
Trade, Competition in the	342
Easily Learned, the	331
Profits in the	340
Work	336
Contract	337
Good and Bad	18
How Low Bids are Secured for	347
In American Houses	16
Responsibility for Bad	19
Poisons in Well Waters	302
Ponds and Streams	307
Positive Disinfectants	317
And Negative Metals	188
Power Required to Raise Water, Table of	234
Practical Plumbers	345
Prescription, a	327
Pressure Due to Head of Water	104
Of Air in Pipes	66
Of Fluids	217
Of Water Due to Head	217
Primers	244
Primitive Methods of Raising Water	235
Privies	86
Country	266
Privy, a Sanitary	267
Privy Council Report on Earth Closets	271
Prony's Formula	222
Protecting Service Pipes from Frost	127
Protective Impurities in Water	163
Salts Formed on Lead	168
Putty Joints in Iron Pipes	49
Pump Barrels, Size of	236
For Outdoor Work, Iron	241
Handles, Leverage of	236
Pumping by Hand	236
By Steam	250
The Labor of	245
Wind Power in	245
Pumps	234
Chain	239
Cistern	239
Durability of Cistern	240
Double-Action Lift and Force	244
Early Forms of	235
For City Houses	244
Iron	239
Lift and Force	242
Motive Power for	235
Pipe Connections for	240
Protecting, from Frost	241
Setting up	237
Vacuum Chambers for	237
Varieties of Hand	245
Windmill, Diameters of	247
Wooden	233
Purification of Water by Contact with Iron	140
Pythogenic Diseases	13
Quicklime	319
Rain, Ocean Salts in	161
Water	161
Solid Matter in	161
Rainfall	145
In United States	229
Ram, Efficiency of the Hydraulic	255
Rams, Capacity of	255
Durability of	257
Fall Required for a Given Duty	254
Hydraulic	253
In Batteries	257
Supply Pipes for	255
Raising Water	234

INDEX.

	Page.
Rarification of Air in Houses	67
Receivers, Foul Pan-Closet	90
Red Lead Joints	48
Reich's Analyses of Lead	150
Renewal of Seals	77
Repairs	342
Ripley, Prof.	171
Roman Cloacæ	87
Rome, Sewer Ventilation in	23
Roofs, Drainage of	229
Roscoe, Prof., on Lead Poisoning in Manchester	156
Rubber Balls in Air Chambers	228
In Water Pipes	134
Rubber Washer Joints	49
Running Traps	113
Rust Joints in Iron Pipes	50
Safe Wastes	54
Safes for Service Pipes, Continuous	113
Safes, Wastes from	55
Safety Valves for Boilers	125
Sags in Service Pipes	112, 124
Salem, Lead Poisoning in	168
Salts, Action of Mixed, on Lead	168, 181, 182
Chemical Law of Action of, upon Lead	164
Conditions Affecting the Solubility of	176
Of Copper, the	214
Of Lead	166
Sanitarian, the Plumber a	334
Sanitary Care of Premises	312
Construction and Drainage of Country Houses	258
Laws of the Jews	23
Policing of Premises	325
Science a Popular Study	5
Work, Practical Benevolence of	6
Public	9
Saturation of Seals by Sewer Gas	76
"Scamping," an Instance of	338
Schuylkill, Carbonates and Sulphates in	172
Seals, Conditions of Security for	78
Depth of	79
Evaporation of	76
In Traps, Dangers which Menace	76
Effect of Changes of Temperature upon	65
Resistance of	66
In Winter, Dangers which Threaten	128
Mistakes Respecting Water	64
Pressures Required to Displace	81
Renewal of	77
Sea Water, Chlorides in	176
Secondary Decomposition	128
Sediment in Traps	82
Service Pipes	19
Continuous Supports for Branch	112
Emptying	112
For Each Floor, Separate	117
In City Houses	104
Should be Accessible	114
Should not be tapped into Water-Closet Basins	99
Sewage, Commercial Value of	88
Contamination of	181
Of American Cities	88
Of English Cities	88
Utilization	88
Sewer Connections, Houses without	338
Gas	23
Ammoniacal Compounds in	26
Analyses of	29
Atmospheric Contamination by	75
Carbonic Acid in	24
Chemical Action of	28
Chemical Composition of	24
Corroding Lead Pipes	74
Diseases Communicated by	35
Effects of, on the Human System	30
In Dwellings, Determining the Presence of	42

	Page.
Sewer Gas in Dwellings	16
Nitrogen in	25
Organic Vapor in	20
Passed through Water Seals	74, 91
Poisoning in Cities	37
Poisoning, Responsibility for	63
Suffocation from	29
Sulphuretted Hydrogen in	25
Transmitted through Brick Walls	75
Ventilation through House Drains	73
Sewers, Breathing of	41
Expansion by Heat of Air in	65
Petroleum in	42
Tide Water in	65
Unventilated	64
Ventilation	24
Impracticable Plans for	40
In Croydon	39
In London	39
In Rome	23
In United States	39
Practical Benefits of	32
Shade Trees, Dangers of	260
Shaded Houses, Unhealthfulness of	260
Ship Closets	95
Modification of	96
Shrinkage of Lead in Cooling	47
Sink Wastes, Obstructions in	45
Traps in	45
Slop Hoppers	103
Smells, Composition of	29
Snow, Water from Melted	146
Soil-Pipe Connections with Sewers	59
Soil Pipes Carried through Foundations	62
Fall Required in	61
For Each Floor	59
Fractures and Broken Joints in	61
Sags in	61
Supports for	61
Sources of Water Supply	299
Specifications, Architects'	200
Loose	338
Sponge Filters	138
Spring and Well Waters	162
Springs	305
Mineral Salts in	306
The Care of	307
Stagnant Water	258
Stay of Proceedings, a	316
Steam Apparatus for Thawing Pipes	132
Pump, An Automatic	253
Pumping, Cost of	253
Pumps	251
For Light Work	251
Stone Drains	60
Strainers for Deep Wells	242
For Slop Hoppers	103
Strains on Pipes	191
Streets of Paris in the Twelfth Century	7
Strength of Pipes	223
Stretch of Lead Pipes	113
Sugar Tests for Organic Matter in Water, the	295
Sulphate of Iron	318
Of Zinc	128
Sulphates	165
Action of, on Lead	172
In Water	172
Sulphide and Sulphate of Iron	173
Of Ammonium Test for Lead, the	199
Of Potassium Test for Lead, the	199
Sulphur and Pitch Joints	50
Waters, Action of, on Lead	173
Sulphuretted Hydrogen	25, 173
Test for Lead, the	198
Sulphuric Acid Test for Lead, the	200
Sulphurous Acid, Disinfectant Properties of	321
Sunlight as a Purifying Agent	260
Supplementary Ventilation for Traps	71, 73
Swarthmore College, Iron Service Pipes in	107

INDEX. 359

Swill, Ash and Garbage Receptacles 313
Syphon, Phenomenon of the 68
Syphoned, Traps which Cannot be 81
Spphoning of Traps in Long Branch Wastes... 71
 Seals in Waste-Pipe Traps 68, 69
Syphons 233

Tacks 113
Tanks and Cisterns 139
Tanks, Cement-Lineu 140
 Cleanliness in 141
 Contamination of Water in 143
 Copper, Brass and Zinc 141
 In City Houses 104, 139
 Iron 140
  Lined with Tinned Copper 140
 Purification of Water in 144
 Sediment in 141
 Size and Capacity of 139
Tanquerel's Observations of Lead Poisoning... 200
Tapping Return Pipe 116
Tests for Zinc in Water 210
Thames Water 162
Thawing Ice in Pipes, Methods of 132
Theory of the Water-Back 122
Thoroughness in Plumbing Work 334
Tile Drains 60
Tin, Action of Nitrates and Nitrites on 211
 Action of Water on 211
 And Lead, Action of Salts on Alloys of... 212
  Electrical Conditions Induced by Contact of 190
 In Well Water, Corrosion of 212
 Lined Iron Pipes for Water Service 109
 Lead Pipes 106
  Corrosion at Joints of 190
  Weight and Strength of 225
 Non-Poisonous Salts of 212
 Tests for 212
 Water Pipes 105
Tinned Copper for Tank Lining 140
 Iron Pipes 109
Tide Water in Sewers 65
Trap Screws 45, 83
Traps 19
 Accumulations in 81
 And Seals 44, 64
 Dangers which Menace Seals in 76
 Dip of 80
 Draught upon 66
 Effect of Changes of Temperature upon Seals in 65
 Experimental Tests of 82
 Flushing of 76
 Forms of 79
 In Kitchen Sink Wastes 45
 In Main Wastes 72
 In Service Pipes, Running 113
 In Waste Pipes 64
 Limit of Utility of 64
 Lodgment of Fæces in 78
 Sediment in 82
 Supplementary Ventilation for 71
 Syphoning of Seals in 68, 69
 Unsealing of 66
 Ventilated 78
 Ventilation of 64
 What Plumbers Find in 83
 Which Cannot be Syphoned 81
Trapping, Examples of Defective 67
 Main Waste Pipes 71
True Disinfectants 321
Tunbridge, Lead Poisoning at 183
Typhoid Conveyed by House Drain 36
 Fever in Cities and Country Districts ... 302
 In Croydon 30
 In Glasgow 74
 In Great Britain 17
 In Lewes 100

Typhoid following Sewers 38
 On a Mountain Top 288

Uncleanliness of Person in the Middle Ages.... 8
Underdrainage of Wet Lands 258
Underground Cisterns 144
Unhealthfulness in Country Districts 288
Unhealthy Houses 13
Unsealing Traps 65
Unskilled Journeymen 332
Unventilated Sewers 64
Utilization of Sewage 83

Vacuum and Safety Valves 125
 Chambers 243
 For Pumps 237
 In Pipes Caused by Flow of Water 71
Valve Closets 93
 Traps 79
Vegetable Acids, Action of, on Lead 178
 In the Merrimac Run 159
Velocity of Flow of Water 218, 219
Ventilated Traps 78
Ventilation as a Protection to Lead Pipes ... 73
 Failure of Attempts at 14
 For Branch Wastes 71
 For Traps, Supplementary 71
 Of Dwellings 14
 Of Roman 23
 Of Sewers 24
  Benefits of 32
  Impracticable Plans for 40
  In Croydon 39
  In London 39
  Through House Drains 73
 Of Traps 64
 Of Waste Pipes 64
  Benefits of 66
 Popular Indifference to 15
Ventilators for Cesspools 277
Venting Pan-Closet Receivers 92
Verdigris 214
Vines and Shade Trees 261
Virus, Power of 316

Waring, George E., Jr 274
Warming Houses 12
Washing Roofs 144
Waste and Soil Pipes 44
Waste of Time by Plumbers 335
 Pipes, Arrangement of 69
  Benefits of Ventilation for 66
  Cast-Iron 40
  Lead 44
  Pressure of Air in 66
  Sizes and Weights of 44
  Trapping and Ventilation of 276
  Traps in Main 72
  Ventilation for Branch 71
  Ventilation of 64
Water, Action of, on Tin 211
 Air in 152
 Ammonia and Organic Matter 214
 Analysis, Amateur 298
 For Lead in 196
 Animal and Vegetable Matter in Raw 179
 Back, Theory of the 122
 Bearing Strata 300
 Carbonates and Sulphates in 172
  In Cochituate 172
  In Croton 172
  In London 172
  In Schuylkill 172
 Carbonic Acid in 297
 Carriage 85
  Substitutes for 85
 Causes of Sickness in 291
 Chemical Action of, on Iron 205
  On Zinc 208

INDEX.

	Page
Water, Chlorides in Sea	176
Chlorine in	293
Collected on Roofs	145
Containing Lead, Advantages of Boiling	200
Corrosion of Copper in	213
Tin in Well	212
Corrosive Impurities in	163
Croton	162
Deceptive Characteristics of	298
Determining Hardness of	292
The Quality of	291
Discharge of	218
Disease Germs in	37
Dissolved Oxygen in	297
Distribution in Houses	115
For Analysis, Concentration of	197
From Boilers	126
Galvanized Iron Tanks	208
Glasgow Cisterns	155
Tanks	139
Gases from the Atmosphere in	306
Hammer, the	227
Head Consumed by Friction of Pipes	220
Impurities in	136
In Distilled	152
In Natural	160
In Spring	306
Lake Cochituate	161
Metallic Salts in	297
Methods of Purifying	308
Nitrates and Chlorides in	163
Nitrates and Nitrites in	293
Organic Acids in Lake and River	158
Impurities in	159
Matter in Waste	275
Physical Properties of	217
Poisoning	100
Potash and Soda in	159
Pressure Due to Head of	217
Pressures	227
Primitive Methods of Raising	235
Purification by Boiling	310
With Neutral Sulphate of Peroxide of Iron	161
Rain	309
Restoring the Flavor of Boiled	310
Sewage Contamination of	181
Solid Matter in Rain	161
Residue of	291
Sources of Nitrates in	174
Organic Contamination of	178
Spring and Well	162, 164
Stagnant	258
Storage of	229
Substances Found in River	162
Sulphates in	172
Supplied to Cities, Organic Matter in	296
Supply in Country Districts	291
Of London	184
Sources of	299
Table of Power Required to Raise	234
Tests for Iron in	207
Zinc in	210
Thames	162
To Calculate the Head of	217
Town Pollution of River	162
Unusual Ingredients in	162
Velocity of Flow of	218
Vitality of Animalculæ in	309
Waste in Cold Weather	129
Weight vs. Pressure	217
Water Closet Connections with Waste Pipes	58
Nuisances	83
Overflows	55
Receivers	90
Requirements of a Good	89
Closets	19
Antiquity of	85
Classification of	89
Deterioration of	102
Disinfection of	93
Effectual Sealing of	102
Evils of Tapping Service Pipes into	100
Examples of Bad	98
Hopper	94
Importance of Abundant Flush for	98
In the Country	269
Inodorous	96
Misplacement of	97
Not Necessarily Objectionable	85
Not Reading Rooms	99
Pan	90
Poisonous Gases from Pan	91
Ship	95
Valve	93
Service, Conditions of a Good	104
Importance of a Constant	104
In Cities	104
ripes, Iron as a Material for	204
Traps	79
Waters, Properties and Composition of Potable	161
Watt, James	251
Weight of Iron Waste Pipes	58
Well Curbs	303
Digging	299
Water, Sources of Contamination in	300
Waters, Poisons in	302
Wells	299
Artesian	162, 303
Braces for Pipes in	242
Driven	305
In Artois	304
In Towns and Villages	301
Location of	300
Woodwork in	303
Wet Cellars	262
Wind Engines, Horse-Power of	250
Wind Power in Pumping	245
Mechanical Application of	249
Windmill Pumps, Duty of	249
Diameter of	247
Windmills, Cost of	248
For Pumping	246
Speed of	247
Wiped Joints	106
Wood Pavements, Unhealthfulness of	33
Wooden Drains	60
Pumps	238
Zinc as a Material for Service Pipes	207
Chemical Action of Water on	208
Compounds	210
Experiments at Columbia College with	209
For Water Conveyance, Dangers of	108
Impurities in	209
Influence of Carbonic Acid and Chlorides on	208
In Water, Corrosion of	209
Tests for	210
Physical Properties of	207
Poisoning	108, 210
Zymotic Diseases	35
Mortality from	17

www.ingramcontent.com/pod-product-compliance
Lightning Source LLC
Chambersburg PA
CBHW020228240426
43672CB00006B/448